I0484640

The R.A.M.S. Library of Alchemy

Volume 6

Four Works of Paracelsus

The Aurora of the Philosophers

The Tincture of the Philosophers

The Treasure of Treasures

The Manual of the Stone of the Philosophers

R.A.M.S. Publishing Company

Four Works of Paracelsus

Philippus Theophrastus Bombast, Paracelsus the Great

The Aurora of the Philosophers

The Tincture of the Philosophers

The Treasure of Treasures

The Manual of the Stone of the Philosophers

Produced by

Restorers of Alchemical Manuscripts Society

R.A.M.S. Publishing Company

R.A.M.S. Publishing Company
117 Rutherford Lane
Stuarts Draft VA 24477

First Edition 2015

ISBN-13 **978-1508632627**
ISBN-10 **1508632626**

Image Processing by Philip N. Wheeler

This book is sold for informational purposes only. Neither the publisher nor the editor shall be held accountable for the use or misuse of the information in this book.

Printed in the United States of America

AVREOLVS PHILIPP? THEOPHRA. PARACELS?

Hic est cui magni mysteria cognita mundi,
Et dare qui potuit de salis arte salem.

117.

Table of Contents

Dedicated to Hans W. Nintzel,

American Alchemist

and

Founder of the

Restorers of Alchemical Manuscripts Society

(R.A.M.S.)

Disclaimer

Liability: The publisher does not warrant or assume any legal liability or responsibility for the accuracy, completeness, or usefulness of any information, apparatus, product, or process disclosed. The publisher makes no representation as to the accuracy or completeness of the contents of this book and specifically disclaims any implied warranty of merchantability or fitness for a particular purpose. No warranty may be created or extended by written sales materials or sales representatives. You should obtain professional consultation where appropriate. The publisher shall not be liable for any loss of profit or other commercial or personal damages, including but not limited to special, incidental, consequential, or other damages.

Introduction

Philip N. Wheeler

This volume of the R.A.M.S. Library contains the following texts attributed to Paracelsus:

The Aurora of the Philosophers

The Tincture of the Philosophers

The Treasure of Treasures

The Manual of the Stone of the Philosophers

The first three texts were distributed at-cost as photocopies by Hans Nintzel to R.A.M.S. members. The fourth text was to be an additional R.A.M.S. work.

That text, The Manual of the Stone of the Philosophers, was sent to me by Hans for retyping. It was in poor shape, apparently copied from a library archive: the first page has the number imprint "323570," which may be a manuscript number. The text includes this note:

```
"Englyshed by J. H. Oxon in 1663;
Mimeographed Jan. 1936."
```

The Aurora
of the
Philosophers

By

Philippus Theophrastus Bombast,

Paracelsus the Great

WHICH HE OTHERWISE CALLS HIS MONARCHIA[1]

---

[1] The work under this title is cited occasionally in other writings of Paracelsus, but is not included in the great folio published at Geneva in 1688. It was first issued at Basle in 1575, and was accompanied with copious annotations in Latin by the editor, Gerard Dorne. This personage was a very persevering collector of the literary remains of Paracelsus, but is not altogether free from the suspicion of having elaborated his original. The Aurora is by some regarded as an instance in point; though no doubt in the main it is a genuine work of the Sage of Hohenheim, yet in some respects it does seem to approximate somewhat closely to previous schools of Alchemy, which can scarcely he regarded as representing the actual standpoint of Paracelsus.

The Aurora of the Philosophers

CHAPTER I.

CONCERNING THE ORIGIN OF THE PHILOSOPHERS' STONE.

ADAM was the first inventor of arts, because he had knowledge of all things as well after the Fall as before[2]. Thence he predicted the world's destruction by water. From this cause, too, it came about that

---

[2] He who created man the same also created science. What has man in any place without labour? When the mandate went forth: Thou shalt live by the sweat of thy brow, there was, as it were, a new creation. When God uttered His fiat the world was made. Art, however, was not then made, nor was the light of Nature. But when Adam was expelled from Paradise, God created for him the light of Nature when He bade him live by the work of his hands. In like manner, He created for Eve her special light when He said to her: In sorrow shalt thou bring forth children. Thus, and there, were these beings made human and earthy that were before like angelicals. ... Thus, by the word were creatures made, and by this same word was also made the light which was necessary to man. ... Hence the interior man followed from the second creation, after the expulsion from Paradise. ... Before the Fall, that cognition which was requisite to man had not begun to develop in him. He received it from the angel when he was cast out of Paradise. ... Man was made complete in the order of the body, but not in the order of the arts. – De Caducis, Par. III.

his successors erected two tables of stone, on which they engraved all natural arts in hieroglyphical characters, in order that their posterity might also become acquainted with this prediction, that so it might be heeded, and provision made in the time of danger. Subsequently, Noah found one of these tables under Mount Araroth, after the Deluge. In this table were described the courses of the upper firmament and of the lower globe, and also of the planets. At length this universal knowledge was divided into several parts, and lessened in its vigour and power. By means of this separation, one man became an astronomer, another a magician, another a cabalist, and a fourth an alchemist. Abraham, that Vulcanic Tubalcain, a consummate astrologer and arithmetician, carried the Art out of the land of Canaan into Egypt, whereupon the Egyptians rose to so great a height and dignity that this wisdom was derived from them by other nations. The patriarch Jacob painted, as it were, the sheep with various colours; and this was done by magic: for in the theology of the Chaldeans, Hebrews, Persians, and Egyptians, they held these arts to be the highest philosophy, to be learnt by their chief nobles and priests. So it was in the time of Moses, when both thc priests and also thc physicians were chosen from among the Magi – the priests for the judgment of what related to health, especially in the knowledge of leprosy. Moses, likewise, was instructed in the Egyptian schools, at the cost and care of Pharaoh's daughter, so that he excelled in all the wisdom and learning of that people. Thus, too, was it with Daniel, who in his youthful days imbibed the learning of the Chaldeans, so that he became a cabalist. Witness his divine predictions and his exposition of those words, "Mene, Mene, Tecelphares". These words can be understood by the prophetic and cabalistic Art. This cabalistic Art was perfectly familiar to, and in constant use by, Moses and the Prophets. The Prophet Elias foretold many things by his cabalistic numbers. So did the Wise Men of old, by this natural and mystical Art, learn to know God rightly. They abode in His laws, and walked in His statutes with great firmness. It is also evident in the Book of Samuel, that the Berelists did not follow the devil's part, but became, by Divine permission, partakers of visions and veritable apparitions, whereof we shall treat more at large in the Book of Supercelestial Things[3]. This gift is granted by the Lord God to those priests who walk in the Divine precepts. It was a custom among the Persians never to admit any one as king unless he were a Wise Man, pre-eminent in

---

[3] No work precisely corresponding to this title is extant among the writings of Paracelsus. The subjects to which reference is made are discussed in the Philosophia Sagax.

reality as well as in name. This is clear from the customary name of their kings; for they were called Wise Men. Such were those Wise Men and Persian Magi who came from the East to seek out thc Lord Jesus, and are called natural priests. The Egyptians, also, having obtained this magic and philosophy from the Chaldeans and Persians, desired that their priests should learn the same wisdom; and they became so fruitful and successful therein that all the neighbouring countries admired them. For this reason Hermes was so truly named Trismegistus, because he was a king, a priest, a prophet, a magician, and a sophist of natural things. Such another was Zoroaster.

CHAPTER II.

WHEREIN IS DECLARED THAT THE GREEKS DREW A LARGE PART OF THEIR LEARNING FROM THE EGYPTIANS; AND HOW IT CAME FROM THEM TO US.

When a son of Noah possessed the third part of the world after the Flood, this Art broke into Chaldaea and Persia, and thence spread into Egypt. The Art having been found out by the superstitious and idolatrous Greeks, some of them who were wiser than the rest betook themselves to the Chaldeans and Egyptians, so that they might draw the same wisdom from their schools. Since, however, the theological study of the law of Moses did not satisfy them, they trusted to their own peculiar genius, and fell away from the right foundation of those natural secrets and arts. This is evident from their fabulous conceptions, and from their errors respecting the doctrine of Moses. It was the custom of the Egyptians to put forward the traditions of that surpassing wisdom only in enigmatical figures and abstruse histories and terms. This was afterwards followed by Homer with marvellous poetical skill; and Pythagoras was also acquainted with it, seeing that he comprised in his writings many things out of the law of Moses and the Old Testament. In like manner, Hippocrates, Thales of Miletus, Anaxagoras, Democritus, and others, did not scruple to fix their minds on the same subject. And yet none of them were practiced in the true Astrology, Geometry, Arithmetic, or Medicine, because their pride prevented this, since they would not admit disciples belonging to other nations than their own. Even when they had got some insight from the Chaldeans and Egyptians, they became more arrogant still than they were before by Nature, and without any diffidence propounded the subject substantially indeed, but mixed with subtle fictions or falsehoods; and then they attempted to elaborate a certain kind of philosophy which descended from them to the Latins. These in their turn, being educated herewith, adorned it with their own doctrines, and by these the philosophy was spread over Europe. Many academies were founded for the propagation of their dogmas and rules, so that the young might be instructed; and this system flourishes with the Germans, and other nations, right down to the present day.

CHAPTER III.

WHAT WAS TAUGHT IN THE SCHOOLS OF THE EGYPTIANS.

The Chaldeans, Persians, and Egyptians had all of them the same knowledge of the secrets of Nature, and also the same religion. It was only the names that differed. The Chaldeans and Persians called their doctrine Sophia and Magic[4]; and the Egyptians, because of the sacrifice, called their wisdom priestcraft. The magic of the Persians, and the theology of the Egyptians, were both of them taught in the schools of old. Though there were many schools and learned men in Arabia, Africa, and Greece, such as Albumazar, Abenzagel, Geber, Rhasis, and Avicenna among the Arabians; and among the Greeks, Machaon, Podalirius, Pythagoras, Anaxagoras, Democritus, Plato, Aristotle, and Rhodianus; still there were different opinions amongst them as to the wisdom of the Egyptian on points wherein they themselves differed, and

---

[4] Before all things it is necessary to have a right understanding of the nature of Celestial Magic. It originates from divine virtue. There is that magic which Moses practised, and there is the maleficent magic of the sorcerers. There are, then, different kinds of Magi. So also there is what is called the Magic of Nature; there is the Celestial Magus; there is the Magus of Faith, that is, one whose faith makes him whole. There is, lastly, the Magus of Perdition. - Philosophia Sagax, Lib. II., c. 6.

whereupon they disagreed with it. For this reason Pythagoras could not be called a wise man, because the Egyptian priestcraft and wisdom were not perpectly taught, although he received therefrom many mysteries and arcana; and that Anaxagoras had received a great many as well, is clear from his discussions on the subject of Sol and its Stone, which he left behind him after his death. Yet he differed in many respects from the Egyptians. Even they would not be called wise men or Magi; but, following Pythagoras, they assumed the name of philosophy: yet they gathered no more than a few gleams like shadows from the magic of the Persians and the Egyptians. But Moses, Abraham, Solomon, Adam, and the wise men that came from the East to Christ, were true Magi, divine sophists and cabalists. Of this art and wisdom the Greeks knew very little or nothing at all; and therefore we shall leave this philosophical wisdom of the Greeks as being a mere speculation, utterly distinct and separate from other true arts and sciences.

CHAPTER IV.

WHAT MAGI THE CHALDEANS, PERSIANS, AND EGYPTIANS WERE.

Many persons have endeavoured to investigate and make use of the secret magic of these wise men; but it has not yet been accomplished. Many even of our own age exalt Trithemius, others Bacon and Agrippa, for magic and the cabala[5] – two things apparently quite distinct – not knowing why they do so. Magic, indeed, is an art and faculty whereby the elementary bodies, their fruits, properties, virtues, and hidden operations are comprehended. But the cabala, by a subtle understanding of the Scriptures, seems to trace out the way to God for men, to shew them how they may act with Him, and prophesy from Him; for the cabala is full of divine mysteries, even as Magic is full of natural secrets. It teaches of and foretells from the nature of things to come as well as of things present, since its operation consists

---

[5] Learn, therefore, Astronomic Magic, which otherwise I call cabalistic. – De Pestilitate, Tract I. This art, formerly called cabalistic, was in the beginning named caballa, and afterwards caballia. It is a species of magic. It was also, but falsely, called Gabanala, by one whose knowledge of the subject was profound. It was of an unknown Ethnic origin, and it passed subsequently to the Chaldaeans and Hebrews, by both of whom it was corrupted. – Philosophia Sagax, Lib. I., s. v. Probatio in Scientiam Nectromantricam.

in knowing the inner constitution of all creatures, of celestial as well as terrestrial bodies: what is latent within them; what are their occult virtues; for what they were originally designed, and with what properties they are endowed. These and the like subjects are the bonds wherewith things celestial are bound up with things of the earth, as may sometimes be seen in their operation even with the bodily eyes. Such a conjunction of celestial influences, whereby the heavenly virtues acted upon inferior bodies, was formerly called by the Magi a Gamahea[6], or the marriage of the celestial powers and properties with elementary bodies. Hence ensued the excellent commixtures of all bodies, celestial and terrestrial, namely, of the sun and planets, likewise vegetables, minerals, and animals.

---

[6] The object which received the influence and exhibited the sign thereof appears to have been termed Gamaheu, Gamahey etc. But the name was chiefly given to certain stones on which various and wonderful images and figures of men and animals have been found naturally depicted, being no work of man, but the result of the providence and counsel of God. – De Imaginibus, c. 7 and c. 13. It is possible, magically, for a man to project his influence into these stones and some other substances. – Ibid., c. 13. But they also have their own inherent virtue, which is indicated by the shape and the special nature of the impression. – Ibid., c. 7. There was also an artificial Gamaheus invented and prepared by the Magi, and this seems to have been more powerful. – De Carduo Angelico.

The devil attempted with his whole force and endeavour to darken this light; nor was he wholly frustrated in his hopes, for he deprived all Greece of it, and, in place thereof, introduced among that people human speculations and simple blasphemies against God and against His Son. Magic, it is true, had its origin in the Divine Ternary and arose from the Trinity of God. For God marked all His creatures with this Ternary and engraved its hieroglyph on them with His own finger. Nothing in the nature of things can be assigned or produced that lacks this magistery of the Divine Ternary, or that does not even ocularly prove it. The creature teaches us to understand and see the Creator Himself, as St. Paul testifies to the Romans. This covenant of the Divine Ternary, diffused throughout the whole substance of things, is indissoluble. By this, also, we have the secrets of all Nature from the four elements. For the Ternary, with the magical Quaternary, produces a perfect Septenary, endowed with many arcana and demonstrated by things which are known. When the Quaternary rests in the Ternary, then arises the Light of the World on the horizon of eternity, and by the assistance of God gives us the whole bond. Here also it refers to the virtues and operations of all creatures, and to their use, since they are stamped and marked with their arcana, signs, characters, and figures, so that there is left in them scarcely the smallest occult point which is not made clear on examination. Then when the Quaternary and the Ternary mount to the Denary is accomplished their retrogression or reduction to unity. Herein is comprised all the occult wisdom of things which God has made plainly manifest to men, both by His word and by the creatures of His hands, so that they may have a true knowledge of them. This shall be made more clear in another place.

CHAPTER V.

CONCERNING THE CHIEF AND SUPREME ESSENCE OF THINGS.

The Magi in their wisdom asserted that all creatures might be brought to one unified substance, which substance they affirm may, by purifications and purgations, attain to so high a degree of subtlety, such divine nature and occult property, as to work wonderful results. For they considered that by returning to the earth, and by a supreme magical separation, a certain perfect substance would come forth, which is at length, by many industrious and prolonged preparations, exalted and raised up above the range of vegetable substances into mineral, above mineral into metallic, and above perfect metallic substances into a perpetual and divine Quintessence[7], including in itself the essence of

---

[7] Man was regarded by Paracelsus as himself in a special manner the true Quintessence. After God had created all the elements, stars, and every other created thing, and had disposed them according to His will, He proceeded, lastly, to the forming of man. He extracted the essence out of the four elements into one mass; He extracted also the essence of wisdom, art, and reason out of the stars, and this twofold essence He congested into one mass: which mass Scripture calls the slime of the earth. From that mass two bodies were made – the sidereal and the elementary. These, according to the light of Nature, are called the quintum esse. The mass was extracted, and therein the firmament and the elements were

all celestial and terrestrial creatures. The Arabs
and Greeks, by the occult characters and
hieroglyphic descriptions of the Persians and the
Egyptians, attained to secret and abstruse
mysteries. When these were obtained and partially
understood they saw with their own eyes, in the
course of experimenting, many wonderful and strange
effects. But since the supercelestial operations lay
more deeply hidden than their capacity could
penetrate, they did not call this a supercelestial
arcanum according to the institution of the Magi,
but the arcanum of the Philosophers' Stone according
to the counsel and judgment of Pythagoras. Whoever
obtained this Stone overshadowed it with various
enigmatical figures, deceptive resemblances,
comparisons, and fictitious titles, so that its
matter might remain occult. Very little or no
knowledge of it therefore can be had from them.

---

condensed. What was extracted from the four after this manner
constituted a fifth. The Quintessence is the nucleus and the
place of the essences and properties of all things in the
universal world. All nature came into the hand of God - all
potency, all property, all essence of the superior and
inferior globe. All these had God joined in His hand, and from
these He formed man according to His image. - Philosophia
Sagax, Lib. I., c. 2.

CHAPTER VI.
CONCERNING THE DIFFERENT ERRORS AS TO ITS DISCOVERY AND KNOWLEDGE.

The philosophers have prefixed most occult names to this matter of the Stone, grounded on mere similitudes. Arnold, observing this, says in his "Rosary" that the greatest difficulty is to find out the material of this Stone; for they have called it vegetable, animal, and mineral, but not according to the literal sense, which is well known to such wise men as have had experience of divine secrets and the miracles of this same Stone. For example, Raymond Lully's "Lunaria" may be cited. This gives flowers of admirable virtues familiar to the philosophers themselves; but it was not the intention of those philosophers that you should think they meant thereby any projection upon metals, or that any such preparations should be made; but the abstruse mind of the philosophers had another intention. In like manner, they called their matter by the name of Martagon, to which they applied an occult alchemical operation; when, notwithstanding that name, it denotes nothing more than a hidden similitude. Moreover, no small error has arisen in the liquid of vegetables, with which a good many have sought to coagulate Mercury[8], and afterwards to convert it with fixatory waters into Luna, since they supposed that he who in this way could coagulate it without the aid of metals would succeed in becoming the chief master. Now, although the liquids of some vegetables do effect this, yet the result is due merely to the resin, fat, and earthy sulphur with which they abound. This attracts to itself the moisture of the Mercury which rises with the substance in the process of coagulation, but without any advantage resulting. I am well assured that no thick and external Sulphur in vegetables is adapted for a perfect projection in Alchemy, as some have found out to their cost. Certain persons have, it is true, coagulated Mercury with the white and milky juice of tittinal, on account of the intense heat which exists therein; and they have called that liquid "Lac Virginis"; yet this is a false basis. The same may be asserted concerning the juice of celandine, although it colours just as though it were endowed with gold. Hence people conceived a vain idea. At a certain fixed time they rooted up this vegetable, from which they sought for a soul or

---

[8] All created things proceed from the coagulated, and after coagulation must go on to resolution. From resolution proceed all procreated things. – De Tartaro (fragment). All bodies of minerals are coagulated by salt. – De Natraralibus Aquis, Lib. III., Tract 2.

quintessence, wherefrom they might make a
coagulating and transmuting tincture. But hence
arose nothing save a foolish error.

CHAPTER VII.

CONCERNING THE ERRORS OF THOSE WHO SEEK THE STONE IN VEGETABLES.

Some alchemists have pressed a juice out of celandine, boiled it to thickness, and put it in the sun, so that it might coagulate into a hard mass, which, being afterwards pounded into a fine black powder, should turn Mercury by projection into Sol. This they also found to be in vain. Others mixed Sal Ammoniac with this powder; others the Colcothar of Vitriol, supposing that they would thus arrive at their desired result. They brought it by their solutions into a yellow water, so that the Sal Ammoniac allowed an entrance of the tincture into the substance of the Mercury. Yet again nothing was accomplished. There are some again who, instead of the abovementioned substances, take the juices of persicaria, bufonaria, dracunculus, the leaves of willow, tithymal, caputia, flammula, and the like, and shut them up in a glass vessel with Mercury for some days, keeping them in ashes. Thus it comes about that the Mercury is turned into ashes, but deceptively and without any result. These people were misled by the vain rumours of the vulgar, who give it out that he who is able to coagulate Mercury without metals has the entire Magistery, as we have said before. Many, too, have extracted salts, oils,

and sulphurs artificially out of vegetables, but
quite in vain. Out of such salts, oils, and sulphurs
no coagulation of Mercury, or perfect projection, or
tincture, can be made. But when the philosophers
compare their matter to a certain golden tree of
seven boughs, they mean that such matter includes
all the seven metals in its sperm, and that in it
these lie hidden. On this account they called their
matter vegetable, because, as in the case of natural
trees, they also in their time produce various
flowers. So, too, the matter of the Stone shews most
beautiful colours in the production of its flowers.
The comparison, also, is apt, because a certain
matter rises out of the philosophical earth, as if
it were a thicket of branches and sprouts: like a
sponge growing on the earth. They say, therefore,
that the fruit of their tree tends towards heaven.
So, then, they put forth that the whole thing hinged
upon natural vegetables, though not as to its
matter, because their stone contains within itself a
body, soul, and spirit, as vegetables do.

CHAPTER VIII.

CONCERNING THOSE WHO HAVE SOUGHT THE STONE IN ANIMALS.

They have also, by a name based only on resemblances, called this matter Lac Virginis, and the Blessed Blood of Rosy Colour, which, nevertheless, suits only the prophets and sons of God. Hence the sophists[9] gathered that this philosophical matter was in the blood of animals or of man. Sometimes, too, because they are nourished by vegetables, others have sought it in hairs, in salt of urine, in rebis; others in hens' eggs, in milk, and in the calx of egg shells, with all of which they thought they would be able to fix Mercury. Some have extracted salt out of foetid urine, supposing that to be the matter of the Stone. Some persons, again, have considered the little stones found in rebis to be the matter. Others have macerated the membranes of eggs in a sharp lixivium, with which they also mixed calcined egg shells as white as snow. To these they have attributed the arcanum of fixation for the transmutation of Mercury. Others, comparing the white of the egg to

---

[9] So acute is the potency of calcined blood, that if it be poured slowly on iron it produces in the first place a whiteness thereon, and then generates rust. – Scholia in Libros de Tartaro. In Lib. II., Tract II.

silver and the yolk to gold, have chosen it for
their matter, mixing with it common salt, sal
ammoniac, and burnt tartar. These they shut up in a
glass vessel, and puri6ed in a Balneum Maris until
the white matter became as red as blood. This,
again, they distilled into a most offensive liquid,
utterly useless for the purpose they had in view.
Others have purified the white and yolk of eggs;
from which has been generated a basilisk. This they
burnt to a deep red powder, and sought to tinge with
it, as they learnt from the treatise of Cardinal
Gilbert. Many, again, have macerated the galls of
oxen, mixed with common salt, and distilled this
into a liquid, with which they moistened the
cementary powders, supposing that, by means of this
Magistery, they would tinge their metals. This they
called by the name of "a part with a part", and
thence came – just nothing. Others have attempted to
transmute tutia by the addition of dragon's blood
and other substances, and also to change copper and
electrum into gold. Others, according to the
Venetian Art, as they call it, take twenty lizard-
like animals, more or less, shut them up in a
vessel, and make them mad with hunger, so that they
may devour one another until only one of them
survives. This one is then fed with filings of
copper or of electrum. They suppose that this
animal, simply by the digestion of his stomach, will

bring about the desired transmutation. Finally, they burn this animal into a red powder, which they thought must be gold; but they were deceived. Others, again, having burned the fishes called truitas (trouts?), have sometimes, upon melting them, found some gold in them; but there is no other reason for it than this: Those fish sometimes in rivers and streams meet with certain small scales and sparks of gold, which they eat. It is seldom, however, that such deceivers are found, and then chiefly in the courts of princes. The matter of the philosophers is not to be sought in animals: this I announce to all. Still, it is evident that the philosophers called their Stone animal, because in their final operations the virtue of this most excellent fiery mystery caused an obscure liquid to exude drop by drop from the matter in their vessels. Hence they predicted that, in the last times, there should come a most pure man upon the earth, by whom the redemption of the world should be brought about; and that this man should send forth bloody drops of a red colour, by means of which he should redeem the world from sin. In the same way, after its own kind, the blood of their Stone freed the leprous metals from their infirmities and contagion. On these grounds, therefore, they supposed they were justified in saying that their Stone was animal.

Concerning this mystery Mercurius speaks as follows to King Calid:

"This mystery it is permitted only to the prophets of God to know. Hence it comes to pass that this Stone is called animal, because in its blood a soul lies hid. It is likewise composed of body, spirit, and soul. For the same reason they called it their microcosm, because it has the likeness of all things in the world, and thence they termed it animal, as Plato named the great world an animal".

CHAPTER IX.

CONCERNING THOSE WHO HAVE SOUGHT THE STONE IN MINERALS.

Hereto are added the many ignorant men who suppose the stone to be three-fold, and to be hidden in a triple genus, namely, vegetable, animal, and mineral. Hence it is that they have sought for it in minerals. Now, this is far from the opinion of the philosophers. They affirm that their stone is uniformly vegetable, animal, and mineral. Now, here note that Nature has distributed its mineral sperm into various kinds, as, for instance, into sulphurs, salts, boraxes, nitres, ammoniacs, alums, arsenics, atraments, vitriols, tutias, haematites, orpiments, realgars, magnesias, cinnabar, antimony, talc, cachymia, marcasites, etc. In all these Nature has not yet attained to our matter; although in some of the species named it displays itself in a wonderful aspect for the transmutation of imperfect metals that are to be brought to perfection. Truly, long experience and practice with fire shew many and various permutations in the matter of minerals, not only from one colour to another, but from one essence to another, and from imperfection to perfection. And, although Nature has, by means of prepared minerals, reached some perfection, yet philosophers will not have it that the matter of the

philosophic stone proceeds out of any of the
minerals, although they say that their stone is
universal. Hence, then, the sophists take occasion
to persecute Mercury himself with various torments,
as with sublimations, coagulations, mercurial
waters, aqua fortis, and the like. All these
erroneous ways should be avoided, together with
other sophistical preparations of minerals, and the
purgations and fixations of spirits and metals.
Wherefore all the preparations of the stone, as of
Geber, Albertus Magnus, and the rest, are
sophistical. Their purgations, cementations,
sublimations, distillations, rectifications,
circulations, putrefactions, conjunctions,
solutions, ascensions, coagulations, calcinations,
and incinerations are utterly profitless, both in
the tripod, in the athanor, in the reverberatory
furnace, in the melting furnace, the accidioneum, in
dung, ashes, sand, or what not; and also in the
cucurbite, the pelican, retort, phial, fixatory, and
the rest. The same opinion must be passed on the
sublimation of Mercury by mineral spirits, for the
white and the red, as by vitriol, saltpetre, alum,
crocuses, etc., concerning all which subjects that
sophist, John de Rupescissa, romances in his
treatise on the White and Red Philosophic Stone.
Taken altogether, these are merely deceitful dreams.
Avoid also the particular sophistry of Geber; for

example, his sevenfold sublimations or
mortifications, and also the revivifications of
Mercury, with his preparations of salts of urine, or
salts made by a sepulchre, all which things are
untrustworthy. Some others have endeavoured to fix
Mercury with: the sulphurs of minerals and metals,
but have been greatly deceived. It is true I have
seen Mercury by this Art, and by such fixations,
brought into a metallic body resembling and
counterfeiting good silver in all respects; but when
brought to the test it has shewn itself to be false.

CHAPTER X.

CONCERNING THOSE WHO HAVE SOUGHT THE STONE AND ALSO PARTICULARS IN MINERALS.

Some sophists have tried to squeeze out a fixed oil from Mercury seven times sublimed and as often dissolved by means of aquafortis. In this way they attempt to bring imperfect metals to perfection: but they have been obliged to relinquish their vain endeavour. Some have purged vitriol seven times by calcination, solution, and coagulation, with the addition of two parts of sal ammoniac, and by sublimation, so that it might be resolved into a white water, to which they have added a third part of quicksilver, that it might be coagulated by water. Then afterwards they have sublimated the Mercury several times from the vitriol and sal ammoniac, so that it became a stone. This stone they affirmed, being conceived of the vitriol, to be the Red Sulphur of the philosophers, with which they have, by means of solutions and coagulations, made some progress in attaining the stone; but in projection it has all come to nothing. Others have coagulated Mercury by water of alum into a hard mass like alum itself; and this they have fruitlessly fixed with fixatory waters. The sophists propose to themselves very many ways of fixing Mercury, but to no purpose, for therein nothing perfect or constant can be had. It is therefore in vain to add minerals thereto by sophistical processes, since by all of them he is stirred up to greater malice, is rendered more lively, and rather brought to greater impurity than to any kind of perfection. So, then, the philosophers' matter is not to be sought from thence. Mercury is somewhat imperfect; and to bring it to perfection will be very difficult, nay, impossible for any sophist. There is nothing therein that can be stirred up or compelled to perfection. Some have taken arsenic several times sublimated, and frequently dissolved with oil of tartar and coagulated. This they have pretended to fix, and by it to turn copper into silver. This, however, is merely a sophistical whitening, for arsenic cannot be fixed[10] unless the operator be an Artist, and knows well its tingeing spirit. Truly in this

[10] One recipe for the fixation of arsenic is as follows: — Take equal parts of arsenic and nitre. Place these in a tigillum, set upon coals so that they may begin to boil and to evaporate. Continue till ebullition and evaporation cease, and the substances shall have settled to the bottom of the vessel like fat melting in a frying-pan; then, for the space of an hour and a half (the longer the better), set it apart to settle. Subsequently pour the compound upon marble, and it will acquire a gold colour. In a damp place it will assume the consistency of a fatty fluid. — De Naturalibus Rebus, c. 9. Again: The fixation of arsenic is performed by salt of urine, after which it is converted by itself into an oil. — Chirurgia Minor, Lib. II.

respect all the philosophers have slept, vainly attempting to accomplish anything thereby. Whoever, therefore, is ignorant as to this spirit, cannot have any hopes of fixing it, or of giving it that power which would make it capable of the virtue of transmutation. So, then, I give notice to all that the whitening of which I have just now spoken is grounded on a false basis, and that by it the copper is deceitfully whitened, but not changed.

Now the sophists have mixed this counterfeit Venus with twice its weight of Luna, and sold it to the goldsmiths and mint-masters, until at last they have transmuted themselves into false coiners – not only those who sold, but those who bought it. Some sophists instead of white arsenic take red, and this has turned out false art; because, however it is prepared, it proves to be nothing but whiteness.

Some, again, have gone further and dealt with common sulphur, which, being so yellow, they have boiled in vinegar, lixivium, or sharpest wines, for a day and a night, until it became white. Then afterwards they sublimated it from common salt and the calx of eggs, repeating the process several times; yet, still, though white, it has been always combustible. Nevertheless, with this they have endeavoured to fix Mercury and to turn it into gold; but in vain. From this, however, comes the most excellent and beautiful cinnabar that I have ever seen. This they propose to fix with the oil of sulphur by cementation and fixation. It does, indeed, give something of an appearance, but still falls short of the desired object. Others have reduced common sulphur to the form of a hepar, boiling it in vinegar with the addition of linseed oil, or laterine oil, or olive oil. They then pour it into a marble mortar, and make it into the form of a hepar, which they have first distilled into a citrine oil with a gentle fire. But they have found to their loss that they could not do anything in the way of transmuting Luna to Sol as they supposed they would be able. As there is an infinite number of metals, so also there is much variety in the preparation of them: I shall not make further mention of these in this place, because each a mould require a special treatise. Beware also of sophisticated oils of vitriol and antimony. Likewise be on your guard against the oils of the metals, perfect or imperfect, as Sol or Luna; because although the operation of these is most potent in the nature of things, yet the true process is known, even at this day, to very few persons. Abstain also from the sophistical preparations of common mercury, arsenic, sulphur, and the like, by sublimation, descension, fixation by vinegar, saltpetre, tartar, vitriol, sal

ammoniac, according to the formulas prescribed in
the books of the sophists. Likewise avoid the
sophisticated tinctures taken from marcasites and
crocus of Mars, and also of that sophistication
called by the name of "a part with a part", and of
fixed Luna and similar trifles. Although they have
some superficial appearance of truth, as the
fixation of Luna by little labour and industry,
still the progress of the preparation is worthless
and weak. Being therefore moved with compassion
towards the well-meaning operators in this art, I
have determined to lay open the whole foundation of
philosophy in three separate arcana, namely, in one
explained by arsenic, in a second by vitriol, and in
a third by antimony; by means of which I will teach
the true projection upon Mercury and upon the
imperfect metals.

CHAPTER XI.

CONCERNING THE TRUE AND PERFECT SPECIAL ARCANUM OF ARSENIC FOR THE WHITE TINCTURE.

Some persons have written that arsenic is compounded of Mercury and, Sulphur, others of earth and water; but most writers say that it is of the nature of Sulphur. But, however that may be, its nature is such that it transmutes red copper into white. It may also be brought to such a perfect state of preparation as to be able to tinge. But this is not done in the way pointed out by such evil sophists as Geber in "The Sum of Perfection", Albertus Magnus, Aristotle the chemist in "The Book of the Perfect Magistery", Rhasis and Polydorus; for those writers, however many they be, are either themselves in error, or else they write falsely out of sheer envy, and put forth receipts whilst not ignorant of the truth. Arsenic contains within itself three natural spirits. The first is volatile, combustible, corrosive, and penetrating all metals. This spirit whitens Venus and after some days renders it spongy. But this artifice relates only to those who practise the caustic art. The second spirit is crystalline and sweet. The third is a tingeing spirit separated from the others before mentioned. True philosophers seek for these three natural properties in arsenic with a view to the perfect projection of the wise men[11]. But those barbers who practise surgery seek after that sweet and crystalline nature separated from the tingeing spirit for use in the cure of wounds, buboes, carbuncles, anthrax, and other similar ulcers which are not curable save by gentle means. As for that tingeing spirit, however, unless the pure be separated from the impure in it, the fixed from the volatile, and the secret tincture from the combustible, it will not in any way succeed according to your wish for projection on Mercury, Venus, or any other imperfect metal. All philosophers have hidden this arcanum as a most excellent mystery. This tingeing spirit, separated from the other two as above, you must join to the spirit of Luna, and digest them together for the space of thirty-two days, or until they have assumed a new body. After it has, on the fortieth natural day, been kindled into flame by the heat of the sun, the spirit appears in a bright whiteness, and is endued with a perfect tingeing arcanum. Then it is

---

[11] Concerning the kinds of arsenic, it is to be noted that there are those which flow forth from their proper mineral or metal, and are called native arsenics. Next there are arsenics out of metals after their kind. Then there are those made by Art through transmutation. White or crystalline arsenic is the best for medicine Yellow and red arsenic are utilised by chemists for investigating the transmutation of metals, in which arsenic has a special efficacy. – De Naturalibus Rebus, c. 9.

at length fit for projection, namely, one part of it upon sixteen parts of an imperfect body, according to the sharpness of the preparation. From thence appears shining and most excellent Luna, as though it had been dug from the bowels of the earth.

CHAPTER XII.

GENERAL INSTRUCTION CONCERNING THE ARCANUM OF VITRIOL AND THE RED TINCTURE TO BE EXTRACTED FROM IT.[12]

Vitriol is a very noble mineral among the rest, and was held always in highest estimation by philosophers, because the Most High God has adorned it with wonderful gifts. They have veiled its arcanum in enigmatical figures like the following: "Thou shalt go to the inner parts of the earth, and by rectification thou shalt find the occult stone, a true medicine". By the earth they understood the Vitriol itself; and by the inner parts of the earth its sweetness and redness, because in the occult part of the Vitriol lies hid a subtle, noble, and most fragrant juice, and a pure oil. The method of its production is not to be approached by calcination or by distillation. For it must not be deprived on any account of its green colour. If it

---

[12] The arcanum of vitriol is the oil of vitriol. Thus: after the aquosity has been removed in coction from vitriol, the spirit is elicited by the application of greater heat. The vitriol then comes over pure in the form of water. This water is combined with the caput mortuum left by the process, and on again separating in a balneum maris, the phlegmatic part passes off, and the oil, or the arcanum of vitriol, remains at the bottom of the vessel. – Ibid.

were, it would at the same time lose its arcanum and its power. Indeed, it should be observed at this point that minerals, and also vegetables and other like things which shew greenness without, contain within themselves an oil red like blood, which is their arcanum. Hence it is clear that the distillations of the druggists are useless, vain, foolish, and of no value, because these people do not know how to extract the bloodlike redness from vegetables. Nature herself is wise, and turns all the waters of vegetables to a lemon colour, and after that into an oil which is very red like blood. The reason why this is so slowly accomplished arises from the too great haste of the ignorant operators who distil it, which causes the greenness to be consumed. They have not learnt to strengthen Nature with their own powers, which is the mode whereby that noble green colour ought to be rectified into redness of itself. An example of this is white wine digesting itself into a lemon colour; and in process of time the green colour of the grape is of itself turned into the red which underlies the coerulean. The greenness therefore of the vegetables and minerals being lost by the incapacity of the operators, the essence also and spirit of the oil and of the balsam, which is noblest among arcana, will also perish.

CHAPTER XIII.

SPECIAL INSTRUCTION CONCERNING THE PROCESS OF VITRIOL FOR THE RED TINCTURE.

Vitriol contains within itself many muddy and viscous imperfections. Therefore its greenness[13] must be often extracted with water, and rectified until it puts off all the impurities of earth. When all these rectifications are finished, take care above all that the matter shall not be exposed to the sun, for this turns its greenness pale, and at the same time absorbs the arcanum. Let it be kept covered up in a warm stove so that no dust may defile it. Afterwards let it be digested in a closed glass vessel for the space of several months, or until different colours and deep redness shew themselves. Still you must not suppose that by this process the redness is sufficiently fixed. It must, in addition, be cleansed from the interior and accidental defilements of the earth, in the following manner: It must be rectified with acetum until the earthy defilement is altogether removed,

---

[13] So long as the viridity or greenness of vitriol subsists therein, it is of a soft quality and substance. But if it be excocted so that it is deprived of its moisture, it is thereby changed into a hard stone from which even fire can be struck. When the moisture is evaporated from vitriol, the sulphur which it contains predominates over the salt, and the vitriol turns red. – De Pestilitate, Tract I.

and the dregs are taken away. This is now the true and best rectification of its tincture, from which the blessed oil is to be extracted. From this tincture, which is carefully enclosed in a glass vessel, an alembic afterwards placed on it and luted so that no spirit may escape, the spirit of this oil must be extracted by distillation over a mild and slow fire. This oil is much pleasanter and sweeter than any aromatic balsam of the drugsellers, being entirely free from all acridity[14]. There will subside in tha bottom of the cucurbite some very white earth, shining and glittering like snow. This

---

[14] The diagnosis of vitriol is concerned with it both in Medicine and Alchemy. In Medicine it is a paramount remedy. In Alchemy it has many additional purposes. The Art of Medicine and Alchemy consists in the preparation of vitriol, for it is worthless in its crude state. It is like unto wood, out of which it is possible to carve anything. Three kinds of oil are extracted from vitriol - a red oil, by distillation in a retort after an alchemistic method, and this is the most acid of all substances, and has also a corrosive quality - also a green and a white oil, distilled from crude vitriol by descension. - De Vitriolo. Nor let it be regarded as absurd that we assign such great virtues to vitriol, for therein resides, secret and hidden, a certain peculiar golden force, not corporeal but spiritual, which excellent and admirable virtue exists in greater potency and certainty therein than it does in gold. When this golden spirit of vitriol is volatilized and separated from its impurities, so that the essence alone remains, it is like unto potable gold. - De Morbis Amentium, Methodus II., c. 1.

keep, and protect from all dust. This same earth is altogether separated from its redness.

Thereupon follows the greatest arcanum, that is to say, the Supercelestial Marriage of the Soul, consummately prepared and washed by the blood of the lamb, with its own splendid, shining, and purified body. This is the true supercelestial marriage by which life is prolonged to the last and predestined day. In this way, then, the soul and spirit of the Vitriol, which are its blood, are joined with its purified body, that they may be for eternity inseparable. Take, therefore, this our foliated earth in a glass phial. Into it pour gradually its own oil. The body will receive and embrace its soul; since the body is affected with extreme desire for the soul, and the soul is most perfectly delighted with the embrace of the body. Place this conjunction in a furnace of arcana, and keep it there for forty days. When these have expired you will have a most absolute oil of wondrous perfection, in which Mercury and any other of the imperfect metals are turned into gold.

Now let us turn our attention to its multiplication. Take the corporal Mercury, in the proportion of two parts; pour it over three parts, equal in weight, of the aforesaid oil, and let them remain together for forty days. By this proportion of weight and this order the multiplication becomes infinite.

CHAPTER XIV.

CONCERNING THE SECRETS AND ARCANA OF ANTIMONY, FOR THE RED TINCTURE, WITH A VIEW TO TRANSMUTATION.

Antimony is the true bath of gold. Philosophers call it the examiner and the stilanx. Poets say that in this bath Vulcan washed Phoebus, and purified him from all dirt and imperfection. It is produced from the purest and noblest Mercury and Sulphur, under the genus of vitriol, in metallic form and brightness. Some philosophers call it the White Lead of the Wise Men, or simply the Lead. Take, therefore, of Antimony, the very best of its kind, as much as you will. Dissolve this in its own aquafortis, and throw it into cold water, adding a little of the crocus of Mars, so that it may sink to the bottom of the vessel as a sediment, for otherwise it does not throw off its dregs. After it has been dissolved in this way it will have acquired supreme beauty. Let it be placed in a glass vessel, closely fastened on all sides with a very thick lute, or else in a stone bocia, and mix with it some calcined tutia, sublimated to the perfect degree of fire. It must be carefully guarded from liquefying, because with too great heat it breaks the glass. From one pound of this Antimony a sublimation is made, perfected for a space of two days. Place this sublimated substance in a phial that it may touch the water with its third part, in a luted vessel, so that the spirit may not escape. Let it be suspended over the tripod of arcana, and let the work be urged on at first with a slow fire equal to the sun's heat at midsummer. Then at length on the tenth day let it be gradually increased. For with too great heat the glass vessels are broken, and sometimes even the furnace goes to pieces. While the vapour is ascending different colours appear. Let the fire be moderated until a red matter is seen. Afterwards dissolve in very sharp Acetum, and throw away the dregs. Let the Acetum be abstracted and let it be again dissolved in common distilled water. This again must be abstracted, and the sediment distilled with a very strong fire in a glass vessel closely shut. The whole body of the Antimony will ascend as a very red oil, like the colour of a ruby, and will flow into the receiver, drop by drop, with a most fragrant smell and a very sweet taste15[15]. This is the supreme arcanum of the philosophers in Antimony,

---

[15] Antimony can be made into a pap with the water of vitriol, and then purified by sal ammoniac, and in this manner there may be obtained from it a thick purple or reddish liquor. This is oil of antimony, and it has many virtues. – Chirurgia Magna, Lib. V. Take three pounds of antimony and as much of sal gemmae. Distil them together in a retort for three natural days, and so you will have a red oil, which has incredible healing power in cases of otherwise incurable wounds. – Chirurgia Minor, Tract II., c. 11.

which they account most highly among the arcana of oils. Then, lastly, let the oil of Sol be made in the following way: Take of the purest Sol as much as you will, and dissolve it in rectified spirit of wine. Let the spirit be abstracted several times, and an equal number of times let it be dissolved again. Let the last solution be kept with the spirit of wine, and circulated for a month. Afterwards let the volatile gold and the spirit of wine be distilled three or four times by means of an alembic, so that it may flow down into the receiver and be brought to its supreme essence. To half an ounce of this dissolved gold let one ounce of the Oil of Antimony be added. This oil embraces it in the heat of the bath, so that it does not easily let it go, even if the spirit of wine be extracted. In this way you will have the supreme mystery and arcanum of Nature, to which scarcely any equal can be assigned in the nature of things. Let these two oils in combination be shut up together in a phial after the manner described, hung on a tripod for a philosophical month, and warmed with a very gentle fire; although, if the fire be regulated in dire proportion this operation is concluded in thirty-one days, and brought to perfection. By this, Mercury and any other imperfect metals acquire the perfection of gold.

CHAPTER XV.

CONCERNING THE PROJECTION TO BE MADE BY THE MYSTERY AND ARCANUM OF ANTIMONY.

No precise weight can be assigned in this work of projection, though the tincture itself may be extracted from a certain subject, in a defined proportion, and with fitting appliances. For instance, that Medicine tinges sometimes thirty, forty, occasionally even sixty, eighty, or a hundred parts of the imperfect metal. So, then, the whole business hinges chiefly on the purification of the Medicine and the industry of the operator, and, next, on the greater; or lesser cleanliness and purity of the imperfect body taken in hand. For instance, one Venus is more pure than another; and hence it happens that no one fixed weight can be specified in projection. This alone is worth noting, that if the operator happens to have taken too much of the tincture, he can correct this mistake by adding more of the imperfect metal. But if there be too much of the subject, so that the powers of the tincture are weakened, this error is easily remedied by a cineritium, or by cementations, or by ablutions in crude Antimony. There is nothing at this stage which need delay the operator; only let him put before himself a fact which has been passed over by the philosophers, and by some studiously veiled, namely, that in projections there must be a revivification, that is to say, an animation of imperfect bodies - nay, so to speak, a spiritualisation; concerning which some have said that their metals are no common ones, since they live and have a soul.

ANIMATION IS PRODUCED IN THE FOLLOWING WAY.

Take of Venus, wrought into small plates, as much as you will, ten, twenty, or forty pounds. Let these be incrusted with a pulse made of arsenic and calcined tartar, and calcined in their own vessel for twenty-four hours. Then at length let the Venus be pulverised, washed, and thoroughly purified. Let the calcination with ablution be repeated three or four times. In this way it is purged and purified from its thick greenness and from its own impure sulphur. You will have to be on your guard against calcinations made with common sulphur. For whatever is good in the metal is spoilt thereby, and what is bad becomes worse. To ten marks of this purged Venus add one of pure Luna. But in order that the work of the Medicine may be accelerated by projection, and may more easily penetrate the imperfect body, and drive out all portions which are opposed to the nature of Luna, this is accomplished by means of a perfect ferment. For the work is defiled by means of an impure Sulphur, so that a cloud is stretched out over the surface of the transmuted substance, or the metal is mixed with the loppings of the Sulphur and may be cast away therewith. But if a projection of a red stone is to be made, with a view to a red transmutation, it must first fall on gold, afterwards on silver, or on some other metal thoroughly purified, as we have directed above. From thence arises the most perfect gold.

CHAPTER XVI.

CONCERNING THE UNIVERSAL MATTER OF THE PHILOSOPHERS' STONE.

After the mortification of vegetables, they are transmuted, by the concurrence of two minerals, such as Sulphur and Salt, into a mineral nature, so that at length they themselves become perfect minerals. So it is that in the mineral burrows and caves of the earth, vegetables are found which, in the long succession of time, and by the continuous heat of sulphur, put off the vegetable nature and assume that of the mineral. This happens, for the most part, where the appropriate nutriment is taken away from vegetables of this kind, so that they are afterwards compelled to derive their nourishment from the sulphur and salts of the earth, until what was before vegetable passes over into a perfect mineral. From this mineral state, too, sometimes a perfect metallic essence arises, and this happens by the progress of one degree into another.

But let us return to the Philosophers' Stone. The matter of this, as certain writers have mentioned, is above all else difficult to discover and abstruse to understand. The method and most certain rule for finding out this, as well as other subjects - what they embrace or are able to effect - is a careful examination of the root and seed by which they come to our knowledge. For this, before all things else, a consideration of principles is absolutely necessary; and also of the manner in which Nature proceeds from imperfection to the end of perfection. Now, for this consideration it is well to have it thoroughly understood from the first that all things created by Nature consist of three primal elements, namely, natural Mercury, Sulphur, and Salt in combination, so that in some substances they are volatile, in others fixed. Wherever corporal Salt is mixed with spiritual Mercury and animated Sulphur into one body, then Nature begins to work, in those subterranean places which serve for her vessels, by means of a separating fire. By this the thick and impure Sulphur is separated from the pure, the earth is segregated from the Salt, and the clouds from the Mercury, while those purer parts are preserved, which Nature again welds together into a pure geogamic body. This operation is esteemed by the Magi as a mixture and conjunction by the uniting of three constituents, body, soul, and spirit. When this union is completed there results from it a pure Mercury. Now if this, when flowing down through its subterranean passages and veins, meets with a chaotic Sulphur, the Mercury is coagulated by it according to the condition of the Sulphur. It is, however, still volatile, so that scarcely in a hundred years is it transformed into a metal. Hence arose the vulgar idea that Mercury and Sulphur are the matter of the metals, as is certainly reported by miners. It is not, however, common Mercury and common Sulphur which are the matter of the metals, but the Mercury and the Sulphur of the philosophers are incorporated and inborn in perfect metals, and in the forms of them, so that they never fly from the fire, nor are they depraved by the force of the corruption caused by the elements. It is true that by the dissolution of this natural mixture our Mercury is subdued, as all the philosophers say. Under this form of words our Mercury comes to be drawn from perfect bodies and from the forces of the earthly planets. This is what Hermes asserts in the following terms: "The Sun and the Moon are the roots of this Art". The Son of Hamuel says that the Stone of the philosophers is water coagulated, namely, in Sol and Luna. From this it is clearer than the sun that the material of the Stone is nothing else but Sol and Luna. This is confirmed by the fact that like produces like. We know that there are only two Stones, the white and the red. There are also two matters of the Stone, Sol and Luna, formed together in a proper marriage, both natural and artificial. Now, as we see that the man or the woman, without the seed of both, cannot generate, in the same way our man, Sol, and his wife, Luna, cannot conceive or

do an thing in the way of generation, without the
seed and sperm of both. Hence the philosophers
gathered that a third thing was necessary, namely,
the animated seed of both, the man and the woman,
without which they judged that the whole of their
work was fruitless and in vain. Such a sperm is
Mercury, which, by the natural conjunction of both
bodies Sol and Luna, receives their nature into
itself in union. Then at length, and not before, the
work is fit for congress, ingress, and generation;
by the masculine and feminine power and virtue.
Hence the philosophers have said that this same
Mercury is composed of body, spirit, and soul, and
that it has assumed the nature and property of all
elements. Therefore, with their most powerful genius
and intellect, they asserted their Stone to be
animal. They even called it their Adam, who carries
his own invisible Eve hidden in his body, from that
moment in which they were united by the power of the
Supreme God, the Maker of all creatures. For this
reason it may be said that the Mercury of the
Philosophers is none other than their most abstruse,
compounded Mercury, and not the common Mercury. So
then they have wisely said to the sages that there
is in Mercury whatever wise men seek. Almadir, the
philosopher, says: "We extract our Mercury from one
perfect body and two perfect natural conditions
incorporated together, which indeed puts forth

externally its perfection, whereby it is able to resist the fire, so that its internal imperfection may be protected by the external perfections". By this passage of the sagacious philosopher is understood the Adamic matter, the limbus of the microcosm[16], and the homogeneous, unique matter of the philosophers. The sayings of these men, which we have before mentioned, are simply golden, and ever to be held in the highest esteem, because they contain nothing superfluous or without force. Summarily, then, the matter of the Philosophers' Stone is none other than a fiery and perfect Mercury

[16] Man himself was created from that which is termed limbus. This limbus contained the potency and nature of all creatures. Hence man himself is called the microcosmus, or world in miniature. – De Generatione Stultorum. Man was fashioned out of the limbus, and this limbus is the universal world. –Paramirum Aliud, Lib. II., c. 2. The limbus was the first matter of man. ... Whosoever knows the limbus knows also what man is. Whatsoever the limbus is, that also is man. – Paramirum Aliud, Lib. IV. There is a dual limbus, man, the lesser limbus, and that Great Limbus from which he was produced. – De Podagra, s. v. de Limbo. The limbus is the seed out of which all creatures are produced and grow, as the tree comes forth from its own special seed. The limbus has its ground in the word of God. – Ibid. The limbus of Adam was haven and earth, water and air. Therefore, man also remains in the limbus, and contains in himself heaven and earth, air and water, and these things he also himself is. – Paragranum Alterum, Tract II.

extracted by Nature and Art; that is, the artificially prepared and true hermaphrodite Adam, and the microcosm: That wisest of the philosophers, Mercurius, making the same statement, called the Stone an orphan. Our Mercury, therefore, is the same which contains in itself all the perfections, force, and virtues of the Sun, which also runs through all the streets and houses of all the planets, and in its own rebirth has acquired the force of things above and things below; to the marriage of which it is to be compared, as is clear from the whiteness and the redness combined in it.

CHAPTER XVII.

CONCERNING THE PREPARATION OF THE MATTER FOR THE PHILOSOPHIC STONE.

What Nature principally requires is that its own philosophic man should be brought into a mercurial substance, so that it may be born into the philosophic Stone. Moreover, it should be remarked that those common preparations of Geber, Albertus Magnus, Thomas Aquinas, Rupescissa, Polydorus, and such men, are nothing more than some particular solutions, sublimations, and calcinations, having no reference to our universal substance, which needs only the most secret fire of the philosophers. Let the fire and Azoth therefore suffice for you. From the fact that the philosophers make mention of certain preparations, such as putrefaction, distillation, sublimation, calcination, coagulation, dealbation, rubification, ceration, fixation, and the like, you should understand that in their universal substance, Nature herself fulfils all the operations in the matter spoken of, and not the operator, only in a philosophical vessel, and with a similar fire, but not common fire. The white and the red spring from one root without any intermediary. It is dissolved by itself, it copulates by itself, grows white, grows red, is made crocus-coloured and black by itself, marries itself and conceives in itself. It is therefore to be decocted, to be baked, to be fused; it ascends, and it descends. All these operations are a single operation and produced by the fire alone. Still, some philosophers, nevertheless, have, by a highly graduated essence of wine, dissolved the body of Sol, and rendered it volatile, so that it should ascend through an alembic, thinking that this is the true volatile matter of the philosophers, though it is not so. And although it be no contemptible arcanum to reduce this perfect metallic body into a volatile, spiritual substance, yet they are wrong in their separation of the elements. This process of the monks, such as Lully, Richard of England, Rupescissa, and the rest, is erroneous. By this process they thought that they were going to separate gold after this fashion into a subtle, spiritual, and elementary power, each by itself, and afterwards by circulation and rectification to combine them again in one – but in vain. For although one element may, in a certain sense, be separated from another, yet, nevertheless, every element separated in this way can again be separated into another element, but these elements cannot afterwards by circulation in a pelican, or by distillation, be again brought back into one; but they always remain a certain volatile matter, and aurum potabile, as they themselves call it. The reason why they could not compass their intention is that Nature refuses to be in this way dragged asunder and separated by man's disjunctions, as by earthly glasses and instruments. She alone knows her own operations and the weights of the elements, the separations, rectifications, and copulations of which she brings about without the aid of any operator or manual artifice, provided only the matter be contained in the secret fire and in its proper occult vessel. The separation of the elements, therefore, is impossible by man. It may appear to take place, but it is not true, whatever may be said by Raymond Lully, and of that famous English golden work which he is falsely supposed to have accomplished. Nature herself has within herself the proper separator, who again joins together what he has put asunder, without the aid of man. She knows best the proportion of every element, which man does not know, however miseading writers romance in their frivolous and false recipes about this volatile gold.

This is the opinion of the philosophers, that when they have put their matter into the more secret fire, and when with a moderated philosophical heat it is cherished on every side, beginning to pass into corruption, it grows black. This operation they term putrefaction, and they call the blackness by

the name of the Crow's Head. The ascent and descent
thereof they term distillation, ascension, and
descension. The exsiccation they call coagulation;
and the dealbation they call calcination; while
because it becomes fluid and soft in the heat they
make mention of ceration. When it ceases to ascend
and remains liquid at the bottom, they say fixation
is present.

In this manner it is the terms of philosophical
operations are to be understood, and not otherwise.

CHAPTER XVIII.

CONCERNING INSTRUMENTS AND THE PHILOSOPHIC VESSEL.

Sham philosophers have misunderstood the occult and secret philosophic vessel, and worse is that which is said by Aristoteles the Alchemist (not the famous Greek Academic Philosopher), giving it out that the matter is to be decocted in a triple vessel. Worst of all is that which is said by another, namely, that the matter in its first separation and first degree requires a metallic vessel; in its second degree of coagulation and dealbation of its earth a glass vessel; and in the third degree, for fixation, an earthen vessel. Nevertheless, hereby the philosophers understand one vessel alone in all the operations up to the perfection of the red stone. Since, then, our matter is our root for the white and the red, necessarily our vessel must be so fashioned that the matter in it may be governed by the heavenly bodies. For invisible celestial influences and the impressions of the stars are in the very first degree necessary for the work: Otherwise it would be impossible for the Oriental, Chaldean, and Egyptian stone to be realised. By this Anaxagoras knew the powers of the whole firmament, and foretold that a great stone would descend from heaven to earth, which actually happened after his death. To the Cabalists our vessel is perfectly well known, because it must be made according to a truly geometrical proportion and measure, and from a definite quadrature of the circle, so that the spirit and the soul of our matter, separated from their body, may be able to raise this vessel with themselves in proportion to the altitude of heaven. If the vessel be wider, narrower, higher, or lower than is fitting, and than the dominating operating spirit and soul desire, the heat of our secret philosophic fire (which is, indeed, very severe), will violently excite the matter and urge it on to excessive operation, so that the vessel is shivered into a thousand pieces, with imminent danger to the body and even the life of the operator. On the other hand, if it be of greater capacity than is required in due proportion for the heat to have effect on the matter, the work will be wasted and thrown away. So, then, our philosophic vessel must be made with the greatest care. What the material of the vessel should be is understood only by those who, in the first solution of our fixed and perfected matter have brought that matter to its own primal quintessence. Enough has been said on this point.

The operator must also very accurately note what, in its first solution, the matter sends forth and rejects from itself.

The method of describing the form of the vessel is difficult. It should be such as Nature requires, and it must be sought out and investigated from every possible source, so that, from the height of the philosophic heaven, elevated above the philosophic earth, it may be able to operate on the fruit of its own earthly body. It should have this form, too, in order that the separation and purification of the elements, when the fire drives one from the other, may be able to be accomplished, and that each may have power to occupy the place to which it adheres; and also that the sun and the other planets may exercise their operations around the elemental earth, while their course in their circuit is neither hindered nor agitated with too swift a motion. In all these particulars which have been mentioned it must have a proper proportion of rotundity and of height.

The instruments for the first purification of mineral bodies are fusing-vessels, bellows, tongs, capels, cupels, tests, cementatory vessels, cineritiums, cucurbites, bocias for aquafortis and aqua regia; and also the appliances which are required for projection at the climax of the work.

CHAPTER XIX.

CONCERNING THE SECRET FIRE OF THE PHILOSOPHERS.

This is a well-known sententious saying of the philosophers, "Let fire and Azoc suffice thee". Fire alone is the whole work and the entire art. Moreover, they who build their fire and keep their vessel in that heat are in error. In vain some have attempted it with the heat of horse dung. By the coal fire, without a medium, they have sublimated their matter, but they have not dissolved it. Others have got their heat from lamps, asserting that this is the secret fire of the philosophers for making their Stone. Some have placed it in a bath, first of all in heaps of ants' eggs; others in juniper ashes. Some have sought the fire in quicklime, in tartar, vitriol, nitre, etc. Others, again, have sought it in boiling water. Thomas Aquinas speaks falsely of this fire, saying that God and the angels cannot do without this fire, but use it daily. What blasphemy is this! Is it not a manifest lie that God is not able to do without the elemental heat of boiling water? All the heats excited by those means which have been mentioned are utterly useless for our work Take care not to be misled by Arnold de Villa Nova, who has written on the subject of the coal fire, for in this matter he will deceive you.

Almadir says that the invisible rays of our fire of themselves suffice. Another cites, as an illustration, that the heavenly heat by its reflections tends to the coagulation and perfection of Mercury, just as by its continual motion it tends to the generation of metals. Again, says this same authority, "Make a fire, vaporous, digesting, as for cooking, continuous, but not volatile or boiling, enclosed, shut off from the air, not burning, but altering and penetrating. Now, in truth, I have mentioned every mode of fire and of exciting heat. If you are a true philosopher you will understand". This is what he says.

Salmanazar remarks: "Ours is a corrosive fire, which brings over our vessel an air like a cloud, in which cloud the rays of this fire are hidden. If this dew of chaos and this moisture of the cloud fail, a mistake has been committed". Again, Almadir says, that unless the fire has warmed our sun with its moisture, by the excrement of the mountain, with a moderate ascent, we shall not be partakers either of the Red or the White Stone.

All these matters shew quite openly to us the occult fire of the wise men. Finally, this is the matter of our fire, namely, that it be kindled by the quiet spirit of sensible fire, which drives upwards, as it were, the heated chaos from the opposite quarter, and above our philosophic matter. This heat, glowing above our vessel, must urge it to the motion of a perfect generation, temperately but continuously, without intermission.

CHAPTER XX.

CONCERNING THE FERMENT OF THE PHILOSOPHERS, AND THE WEIGHT.

Philosophers have laboured greatly in the art of ferments and of fermentations, which seems important above all others. With reference thereto some have made a vow to God and to the philosophers that they would never divulge its arcanum by similitudes or by parables.

Nevertheless, Hermes, the father of all philosophers, in the "Book of the Seven Treatises", most clearly discloses the secret of ferments, saying that they consist only of their own paste; and more at length he says that the ferment whitens the confection, hinders combustion, altogether retards the flux of the tincture, consoles bodies, and amplifies unions. He says, also, that this is the key and the end of the work, concluding that the ferment is nothing but paste, as that of the sun is nothing but sun, and that of the moon nothing but moon. Others affirm that the ferment is the soul, and if this be not rightly prepared from the magistery, it effects nothing. Some zealots of this Art seek the Art in common sulphur, arsenic, tutia, auripigment, vitriol, etc., but in vain; since the substance which is sought is the same as that from which it has to be drawn forth. It should be remarked, therefore, that fermentations of this kind do not succeed according to the wishes of the zealots in the way they desire, but, as is clear from what has been said above, simply in the way of natural successes.

But, to come at length to the weight; this must be noted in two ways. The first is natural, the second artificial. The natural attains its result in the earth by Nature and concordance. Of this, Arnold says: If more or less earth than Nature requires be added, the soul is suffocated, and no result is perceived, nor any fixation. It is the same with the water. If more or less of this bc taken it will bring a corresponding loss. A superfluity renders the matter unduly moist, and a deficiency makes it too dry and too hard. If there be over much air present, it is too strongly impressed on the tincture; if there be too little, the body will turn out pallid. In the same way, if the fire be too strong, the matter is burnt up; if it be too slack, it has not the power of drying, nor of dissolving or heating the other elements. In these things elemental heat consists.

Artificial weight is quite occult. It is comprised in the magical art of ponderations. Between the spirit, soul, and body, say the philosophers, weight consists of Sulphur as the director of the work; for the soul strongly desires Sulphur, and necessarily observes it by reason of its weight.

You can understand it thus: Our matter is united to a red fixed Sulphur, to which a third part of the regimen has been entrusted, even to the ultimate degree, so that it may perfect to infinity the operation of the Stone, may remain therewith together with its fire, and may consist of a weight equal to the matter itself, in and through all, without variation of any degree. Therefore, after the matter has been adapted and mixed in its proportionate weight, it should be closely shut up with its seal in the vessel of the philosophers, and committed to the secret fire. In this the Philosophic Sun will rise and surge up, and will illuminate all things that have been looking for his light, expecting it with highest hope.

In these few words we will conclude the arcanum of the Stone, an arcanum which is in no way maimed or defective, for which we give God undying thanks. Now have we opened to you our treasure, which is not to be paid for by the riches of the whole world.

HERE ENDS THE AURORA OF THE PHILOSOPHERS.

The Book

Concerning The Tincture

Of The Philosophers

by

Philippus Theophrastus Bombast,
Paracelsus the Great

From "Paracelsus his Archidoxis"

Faithfully and plainly Englished, and Published
by, J.H. Oxon. London, Printed for W.S. and are
to be sold by Thomas Brewster at the Three
Bibles in Pauls Church-yard. 1660.

The Book Concerning The Tincture Of The Philosophers

by

Paracelsus

THE BOOK CONCERNING THE TINCTURE OF THE PHILOSOPHERS
WRITTEN AGAINST THOSE SOPHISTS BORN SINCE THE
DELUGE, IN THE AGE OF OUR LORD JESUS CHRIST, THE SON
OF GOD;

By

PHILIPPUS THEOPHRASTUS BOMBAST, of HOHENHEIM,
Philosopher of the Monarchia, Prince of Spagyrists,
Chief Astronomer, Surpassing Physician, and
Trismegistus of Mechanical Arcana.

The Book
Concerning The Tincture
Of The Philosophers
by
Paracelsus

PREFACE

SINCE you, O Sophist, everywhere abuse me with such fatuous and mendacious words, on the ground that being sprung from rude Helvetia I can understand and know nothing: and also because being a duly qualified physician I still wander from one district to another; therefore I have proposed by means of this treatise to disclose to the ignorant and inexperienced: what good arts existed in the first age; what my art avails against you and yours against me; what should be thought of each, and how my posterity in this age of grace will imitate me. Look at Hermes, Archelaus, and others in the first age: see what Spagyrists and what Philosophers then existed. By this they testify that their enemies, who are your patrons, O Sophist, at the present time are but mere empty forms and idols. Although this would not be attested by those who are falsely considered your authentic fathers

and saints, yet the ancient Emerald Table shews
more art and experience in Philosophy, Alchemy,
Magic, and the like, than could ever be taught
by you and your crowd of followers. If you do
not yet understand, from the aforesaid facts,
what and how great treasures these are, tell me
why no prince or king was ever able to subdue
the Egyptians. Then tell me why the Emperor
Diocletian ordered all the Spagyric books to be
burnt (so far as he could lay his hands upon
them). Unless the contents of those books had
been known, they would have been obliged to bear
still his intolerable yoke, - a yoke, O Sophist,
which shall one day be put upon the neck of
yourself and your colleagues.

From the middle of this age the Monarchy of all
the Arts has been at length derived and
conferred on me, Theophrastus Paracelsus, Prince
of Philosophy and of Medicine. For this purpose
I have been chosen by God to extinguish and blot
out all the phantasies of elaborate and false
works, of delusive and presumptuous words, be
they the words of Aristotle, Galen, Avicenna,
Mesva, or the dogmas of any among their
followers. My theory, proceeding as it does from
the light of Nature, can never, through its
consistency, pass away or be changed: but in the

fifty-eighth year after its millennium and a half it will then begin to flourish. The practice at the same time following upon the theory will be proved by wonderful and incredible signs, so as to be open to mechanics and common people, and they will thoroughly understand how firm and immovable is that Paracelsic Art against the triflings of the Sophists: though meanwhile that sophistical science has to have its ineptitude propped up and fortified by papal and imperial privileges. In that I am esteemed by you a mendicant and vagabond sophist, the Danube and the Rhine will answer that accusation, though I hold my tongue. Those calumnies of yours falsely devised against me have often displeased many courts and princes, many imperial cities, the knightly order, and the nobility. I have a treasure hidden in a certain city called Weinden, belonging to Forum Julii, at an inn, - a treasure which neither you, Leo of Rome, nor you, Charles the German, could purchase with all your substance. Although the signed star has been applied to the arcanum of your names, it is known to none but the sons of the divine Spagyric Art. So then, you wormy and lousy Sophist, since you deem the monarch of arcana a mere ignorant, fatuous, and prodigal quack, now, in this mid age, I determine in my present treatise to disclose the honourable course of procedure in these matters, the virtues and preparation of the celebrated Tincture of the Philosophers for the use and honour of all who love the truth, and in order that all who despise the true arts may be reduced to poverty. By this arcanum the last age shall be illuminated clearly and compensated for all its losses by the gift of grace and the reward of the spirit of truth, so that since the beginning of the world no similar germination of the intelligence and of wisdom shall ever have been heard of. In the meantime, vice will not be able to suppress the good, nor will the resources of those vicious persons, many though they be,

cause any loss to the upright.

CHAPTER I

I, PHILIPPUS Theophrastus Paracelsus Bombast, say that, by Divine grace, many ways have been sought to the Tincture of the Philosophers, which finally all came to the same scope and end. Hermes Trismegistus, the Egyptian, approached this task in his own method. Orus, the Greek, observed the same process. Hali, the Arabian, remained firm in his order. But Albertus Magnus, the German, followed also a lengthy process. Each one of these advanced in proportion to his own method; nevertheless, they all arrive at one and the same end, at a long life, so much desired by the philosophers, and also at an honourable sustenance and means of preserving that life in this Valley of Misery. Now at this time, I, Theophrastus Paracelsus Bombast, Monarch of the Arcana, am endowed by God with special gifts for this end, that every searcher after this supreme philosophic work may be forced to imitate and to follow me, be he Italian, Pole, Gaul, German, or whatsoever or whosoever he be. Come hither after me, all you philosophers, astronomers, and spagyrists, of however lofty a name ye may be, I will show and open to you, Alchemists and Doctors, who are exalted by me with the most consummate labours, this corporeal regeneration. I will teach you the tincture, the arcanum[17], the quintessence, wherein lie hid the foundations of all mysteries and of all works. For every person may and ought to believe in another only in those matters, which he has tried by fire. If anyone shall have brought forward anything contrary to this method of experimentation in the Spagyric Art or in Medicine, there is no reason for your belief in him, since, experimentally, through the agency of fire, the true is separated from the false. The light of Nature indeed is created in this way, that by means thereof the proof or trial of everything may appear, but only to those who walk in this light. With this light we will teach, by the very best methods of demonstration, that all those who before me have approached this so difficult province with their

---

[17] The Arcanum of a substance is not the virtue (*virtus*) but the essence (*vir*) and the potency (*potentia*), and is stronger than the virtue; nevertheless, an old error of the doctors conferred the name of virtues upon the potential essences. - *Paramirum*, Lib. IV. Many things are elsewhere set forth concerning the Quintessence, but what is described is really a separation or extraction of the pure from the impure, not a true quintessence, and it is more correctly termed an Arcanum. -*Explicatio Totius Astronomiae*.

own fancies and acute speculations have, to their own loss, incurred the danger of their foolishness. On which account, from my standpoint, many rustics have been ennobled; but, on the other hand, through the speculative and opinionative art of these many nobles have been changed into clowns, and since they carried golden mountains in their head before they had put their hand to the fire. First of all, then, there must be learnt - digestions, distillations, sublimations, reverberations, extractions, solutions, coagulations, fermentations, fixations, and every instrument which is requisite for this work must be mastered by experience, such as glass vessels, cucurbites, circulators, vessels of Hermes, earthen vessels, baths, blast-furnaces, reverberatories, and instruments of like kind, also marble, coals, and tongs. Thus at length you will be able to profit in Alchemy and in Medicine.

But so long as, relying on fancy and opinion, you cleave to your fictitious books, you are fitted and predestinated for no one of these things.

CHAPTER II

CONCERNING THE DEFINITION OF THE SUBJECT AND MATTER OF THE TINCTURE OF THE PHILOSOPHERS

Before I come, then, to the process of the Tincture, it is needful that I open to you the subject thereof: for, up to the present time, this has always been kept in a specially occult way by the lovers of truth. So, then, the matter of the Tincture (when you understand me in a Spagyrical sense) is a certain thing which, by the art of Vulcan[18], passes out of three essences

---

[18] The office of Vulcan is the separation of the good from the bad. So the Art of Vulcan, which is Alchemy, is like unto death, by which the eternal and the temporal are divided one from another. So also this art might be called the death of things. - *De Morbis Metallicis*, Lib. I., Tract III., c. 1. Vulcan is an astral and not a corporal fabricator. - *De Caduco Matricis*, Par. VI. The artist working in metals and other minerals transforms them into other colours, and in so doing his operation is like that of the heaven itself. For as the artist excocts by means of Vulcan, or the igneous element, so heaven performs the work of coction through the Sun. The Sun, therefore, is the Vulcan of heaven accomplishing coction in the earth. -

into one essence, or it may remain. But, that I may give it its proper name, according to the use of the ancients, though it is called by many the Red Lion, still it is known by few. This, by the aid of Nature and the skill of the Artist himself, can be transmuted into a White Eagle, so that out of one two are produced; and beyond this the brightness of gold does not shine so much for the Spagyrist as do these two when kept in one. Now, if you do not understand the use of the Cabalists and the old astronomers, you are not born by God for the Spagyric art, or chosen by Nature for the work of Vulcan, or created to open your mouth concerning Alchemical Arts. The matter of the Tincture, then, is a very great pearl and a most precious treasure, and the noblest thing next to the manifestation of the Most High and the consideration of men which can exist upon earth. This is the Lili of Alchemy and of Medicine, which the philosophers have so diligently sought after, but, through the failure of entire knowledge and complete

---

De Icteritiis. Vulcan is the fabricator and architect of all things, nor is his habitation in heaven only, that is, in the firmament, but equally in all the other elements. - *Lib. Meteorum*, c. 4. Where the three prime principles are wanting, there also the igneous essence is absent. The Igneous Vulcan is nothing else but Sulphur, Sal Nitrum, and Mercury. - *Ibid.*, c.5.

preparation, they have not progressed to the perfect end thereof. By means of their investigations and experiments, only the initial stage of the Tincture has been given to us; but the true foundation, which my colleagues must imitate, has been left for me, so that no one should mingle their shadows with our good intentions. I, by right after my long experiences, correct the Spagyrists, and separate the false or the erroneous from the true, since, by long investigations, I have found reasons why I should be able justly to blame and to change diverse things. If, indeed, I had found out experiments of the ancients better than my own, I should scarcely have taken up such great labours as, for the sake, the utility, and the advantage of all good Alchemists, I have undergone willingly. Since, then, the subject of the Tincture has been sufficiently declared, so that it scarcely could or ought to be exceeded in fidelity between two brothers, I approach its preparation, and after I have laid down the experiences of the first age, I wish to add my own inventions; to which at last the Age of Grace will by-and-by give its adhesion, whichever of the patriarchs, O Sophist, you, in the meantime, shall have made leaders.

CHAPTER III

CONCERNING THE PROCESS OF THE ANCIENTS FOR THE TINCTURE OF THE PHILOSOPHERS, AND A MORE COMPENDIOUS METHOD BY PARACELSUS

The old Spagyrists putrefied Lili for a philosophical month, and afterwards distilled therefrom the moist spirits, until at length the dry spirits were elevated. They again imbued the caput mortuum with moist spirits, and drew them off from it frequently by distillation until the dry spirits were all elevated. Then afterwards they united the moisture that had been drawn off and the dry spirits by means of a pelican, three or four times, until the whole Lili remained dry at the bottom. Although early experience gave this process before fixation, none the less our ancestors often attained a perfect realization of their wish by this method. They would, however, have had a shorter way of arriving at the treasure of the Red Lion if they had learnt the agreement of Astronomy with Alchemy, as I have demonstrated it in the Apocalypse of Hermes[19].

But since every day (as Christ says for the consolation of the faithful) has its own peculiar care, the labour for the Spagyrists before my times has been great and heavy; but this, by the help of the Holy Spirit flowing into us, will, in this last age, be lightened and made clear by my theory and practice, for all those who constantly persevere in their work with patience. For I have tested the properties of Nature, its essences and conditions, and I know its conjunction and resolution, which are the highest and greatest gift for a philosopher, and never understood by the sophists up to this time. When, therefore, the earliest age gave the first experience of the Tincture, the Spagyrists made two things out of one simple. But when afterwards, in the Middle Age, this invention had died out, their successors by diligent scrutiny afterwards came upon the two names of this simple, and they named it with one word, namely, Lili, as being the subject of the Tincture. At length the imitators of Nature putrefied this matter at its proper period just like the seed in the earth, since before this corruption nothing could be born from it, nor any arcanum break forth from it. Afterwards they drew off the moist spirits from the matter,

[19] For extensive notes please see Appendix A.

until at length, by the violence of the fire, the dry were also equally sublimated, so that, in this way, just as the rustic does at the proper time of year, they might come to maturity as one after another is wont to ascend and to fall away. Lastly, as after the spring comes summer, they incorporated those fruits and dry spirits, and brought the Magistery of the Tincture to such a point that it came to the harvest, and laid itself out for ripening.

CHAPTER IV

CONCERNING THE PROCESS FOR THE TINCTURE OF THE PHILOSOPHERS, AS IT IS SHORTENED BY PARACELSUS

The ancient Spagyrists would not have required such lengthened labour and such wearisome repetition if they had learnt and practised their work in my school. They would have obtained their wish just as well, with far less expense and labour. But at this time, when Theophrastus Paracelsus has arrived as the Monarch of Arcana, the opportunity is at hand for finding out those things which were occult to all Spagyrists before me. Wherefore I say, Take only the rose-coloured blood from the Lion and the gluten from the Eagle. When you have mixed these, coagulate them according to the old process, and you will have the Tincture of the Philosophers, which an infinite number have sought after and very few have found. Whether you will or not, sophist, this Magistery is in Nature itself, a wonderful thing of God above Nature, and a most precious treasure in this Valley of Sorrows. If you look at it from without it seems a paltry thing to transmute another into something far more noble than it was before. But you must, nevertheless, allow this, and confess that it is a miracle produced by the Spagyrist, who by the art of his preparation corrupts a visible body which is externally vile, from which he excites another most noble and most precious essence. If you, in like manner, have learnt anything from the light of Aristotle, or from us, or from the rules of Serapio, come forth, and bring that knowledge experimentally to light. Preserve now the right of the Schools, as becomes a lover of honour and a doctor. But if you know nothing and can do nothing, why do you despise me as though I were an irrational Helvetian cow, and inveigh against me as a wandering vagabond? Art is a second Nature and a universe of its own, as experience witnesses, and demonstrates against you and your idols. Sometimes, therefore, the Alchemist compounds certain simples, which he afterwards corrupts according to his need, and prepares thence another thing. For thus very often out of many things one is made, which effects more than Nature of herself can do, as in Gastaynum it is perfectly well known that Venus is produced from Saturn; in Carinthia, Luna out of Venus; and in Hungary, Sol out of Luna; to pass over in silence for the time being the transmutations of other natural objects, which were well known to the Magi, and more wonderfully than Ovid narrates in his Metamorphoses do they come to the light. That you may rightly understand me, seek your Lion in the East, and your Eagle in the South, for this our work which has been undertaken. You will not find better instruments than Hungary and Istria produce. But if you desire to lead from unity by duality in trinity with equal permutation of each, then you should direct your journey to the South; so in Cyprus shall you gain all your desire, concerning which we must not dilate more profusely than we have done at present. There are still many more of these arcana which exhibit transmutations, though they are known to few. And although these may by the Lord God be made manifest to anyone, still, the rumour of this Art does not on that account at once break forth, but the Almighty gives therewith the understanding how to conceal these and other like arts even to the coming of Elias the Artist, at which time there shall be nothing so occult that it shall not be revealed. You also see with your eyes (though there is no need to speak of these things, which may be taken derisively by some) that in the fire of Sulphur is a great tincture for gems, which, indeed, exalts them to a loftier degree than Nature by herself could do. But this gradation of metals and gems shall be omitted by me in this place, since I have written sufficiently about it in my Secret of Secrets, in my book on the Vexations of Alchemists, and abundantly elsewhere. As I have begun the process of our ancestors with the Tincture of the Philosophers, I will now perfectly conclude it.

CHAPTER V

CONCERNING THE CONCLUSION OF THE PROCESS OF THE ANCIENTS, MADE BY PARACELSUS

Lastly, the ancient Spagyrists having placed Lili in a pelican and dried it, fixed it by means of a regulated increase of the fire, continued so long until from blackness, by permutation into all the colours, it became red as blood, and therewith assumed the condition of a salamander. Rightly, indeed, did they proceed with such labour, and in the same way it is right and becoming that everyone should proceed who seeks this pearl. It will be very difficult for me to make this clearer to you unless you shall have learnt in the School of the Alchemists to observe the degrees of the fire, and also to change your vessels. For then at length you will see that soon after your Lili shall have become heated in the Philosophic Egg, it becomes, with wonderful appearances, blacker than the crow; afterwards, in succession of time, whiter than the swan; and at last, passing through a yellow colour, it turns out more red

than any blood. Seek, seek, says the first
Spagyrist, and you shall find; knock, and it
shall be opened unto you. It would be impious
and indecorous to put food in the mouth of a
perfidious bird. Let her rather fly to it, even
as I, with others before me, have been compelled
to do. But follow true Art; for this will lead
you to its perfect knowledge. It is not possible
that anything should here be set down more fully
or more clearly than I have before spoken. Let
your Pharisaical schools teach you what they
will from their unstable and slippery
foundation, which reaches not its end or its
aim. When at length you shall have been taught
as accurately as possible the Alchemistic Art,
nothing in the nature of things shall then at
length be so difficult which cannot be made
manifest to you by the aid of this Art. Nature,
indeed, herself does not bring forth anything
into the light which is advanced to its highest
perfection, as can be seen in this place from
the unity, or the union, of our duality. But a
man ought by Spagyric preparations to lead it
thither where it was ordained by Nature. Let
this have been sufficiently said by me,
concerning the process of the ancients and my
correction of the Tincture of the Philosophers,
so far as relates to its preparation.

Moreover, since now we have that treasure of the Egyptians in our hands, it remains that we turn it to our use: and this is offered to us by the Spagyric Magistery in two ways. According to the former mode it can be applied for the renewing of the body; according to the latter it is to be used for the transmutation of metals. Since, then, I, Theophrastus Paracelsus, have tried each of them in different ways, I am willing to put them forward and to describe them according to the signs indeed of the work, and as in experience and proof they appeared to me better and more perfectly.

CHAPTER VI

CONCERNING THE TRANSMUTATION OF METALS BY THE PERFECTION OF MEDICINE

If the Tincture of the Philosophers is to be used for transmutation, a pound of it must be projected on a thousand pounds of melted Sol. Then, at length, will a Medicine have been prepared for transmuting the leprous moisture of the metals. This work is a wonderful one in the light of Nature, namely, that by the Magistery, or the operation of the Spagyrist, a metal, which formerly existed, should perish, and another be produced. This fact has rendered that same Aristotle, with his ill-founded philosophy, fatuous. For truly, when the rustics in Hungary cast iron at the proper season into a certain fountain, commonly called Zifferbrunnen, it is consumed into rust, and when this is liquefied with a blast-fire, it soon exists as pure Venus, and never more returns to iron. Similarly, in the mountain commonly called Kuttenberg, they obtain a lixivium out of marcasites, in which iron is forthwith turned into Venus of a high grade, and more malleable than the other produced by Nature. These things, and more like them, are known to simple men rather than to sophists, namely, those which turn one appearance of a metal into another. And these things, moreover, through the remarkable contempt of the ignorant, and partly, too, on account of the just envy of the artificers, remain almost hidden. But I myself, in Istria, have often brought Venus to more than twenty-four (al. 38) degrees, so that the colour of Sol could not mount higher, consisting of Antimony or or Quartal, which Venus I used in all respects as other kinds.

But though the old artists were very desirous of this arcanum, and sought it with the greatest diligence, nevertheless, very few could bring it by means of a perfect preparation to its end. For the transmutation of an inferior metal into a superior one brings with it many difficulties and obstacles, as the change of Jove into Luna, or Venus into Sol. Perhaps on account of their sins God willed that the Magnalia of Nature should be hidden from many men. For sometimes, when this Tincture has been prepared by artists, and they were not able to reduce their projection to work its effects, it happened that, by their carelessness and bad guardianship, this was eaten up by fowls, whose feathers thereupon fell off, and, as I myself have seen, grew again. In this way transmutation, through its abuse from the carelessness of the artists, came into Medicine and Alchemy. For when they were unable to use the Tincture according to their desire, they converted the same to the renovation of men, as shall be heard more at large in the following chapter.

CHAPTER VII

CONCERNING THE RENOVATION OF MEN

Some of the first and primitive philosophers of Egypt have lived by means of this Tincture for a hundred and fifty years. The life of many, too, has been extended and prolonged to several centuries, as is most clearly shewn in different histories, though it seems scarcely credible to any one. For its power is so remarkable that it extends the life of the body beyond what is possible to its congenital nature, and keeps it so firmly in that condition that it lives on in safety from all infirmities. And although, indeed, the body at length comes to old age, nevertheless, it still appears as though it were established in its primal youth.

So, then, the Tincture of the Philosophers is a Universal Medicine, and consumes all diseases, by whatsoever name they are called, just like an invisible fire. The dose is very small, but its effect is most powerful. By means thereof I have cured the leprosy, venereal disease, dropsy, the falling sickness, colic, scab, and similar afflictions; also lupus, cancer, noli-me-

tangere, fistulas, and the whole race of
internal diseases, more surely than one could
believe. Of this fact Germany, France, Italy,
Poland, Bohemia, etc., will afford the most
ample evidence.

Now, Sophist, look at Theophrastus Paracelsus.
How can your Apollo, Machaon, and Hippocrates
stand against me? This is the Catholicum of the
Philosophers, by which all these philosophers
have attained long life for resisting diseases,
and they have attained this end entirely and
most effectually, and so, according to their
judgment, they named it The Tincture of the
Philosophers. For what can there be in the whole
range of medicine greater than such purgation of
the body, by means whereof all superfluity is
radically removed from it and transmuted? For
when the seed is once made sound all else is
perfected. What avails the ill-founded purgation
of the sophists since it removes nothing as it
ought? This, therefore, is the most excellent
foundation of a true physician, the regeneration
of the nature, and the restoration of youth.
After this, the new essence itself drives out
all that is opposed to it. To effect this
regeneration, the powers and virtues of the
Tincture of the Philosophers were miraculously

discovered, and up to this time have been used
in secret and kept concealed by true Spagyrists.

APPENDIX A

The Book of the Revelation of Hermes,
interpreted by Theophrastus Paracelsus,
concerning the Supreme Secret of the World,
seems to have been first brought to light by
Benedictus Figulus, and appeared as a *piece de
résistance* in his "Golden and Blessed Casket of
Nature's Marvels", of which an English
translation has been very recently published.
("A Golden and Blessed Casket of Nature's
Marvels". By Benedictus Figulus. Now first done
into English from the German original published
at Frankfort in the year 1608. London: James
Elliott and Co. 8vo., 1893). Among the many
writings which have been fabulously attributed
to Hermes, there does not seem to be any record
of an apocalypse, and it is impossible to say
what forged document may have been the subject
of interpretation by Paracelsus. As the
collection of Figulus is now so readily
accessible, it is somewhat superfluous to
reproduce the treatise here, but since this
translation claims to include everything written
by the physician of Hohenheim on the subject of
Alchemy and the Universal Medicine, it is
appended at this point. It should be premised
that Benedictus Figulus complains bitterly of

the mutilation and perversion to which the works of Paracelsus were subjected, and the Revelation of Hermes seems in many parts to betray another hand, especially in its quotation of authorities who are not countenanced by its reputed author. Hermes, Plato, Aristotle, and other philosophers, flourishing at different times, who have introduced the Arts, and more especially have explored the secrets of inferior Creation, all these have eagerly sought a means whereby man's body might be preserved from decay and become endued with immortality. To them it was answered that there is nothing which might deliver the mortal body from death; but that there is One Thing which may postpone decay, renew youth; and prolong short human life (as with the patriarchs). For death was laid as a punishment upon our first parents, Adam and Eve, and will never depart from all their descendants. Therefore, the above philosophers, and many others, have sought this One Thing with great labour, and have found that that which preserves the human body from corruption, and prolongs life, conducts itself, with respect to other elements, as it were like the Heavens; from which they understood that the Heavens are a substance above the Four Elements. And just as the Heavens, with respect to the other elements, are held to be the fifth substance (for they are indestructible, stable, and suffer no foreign admixture), so also this One Thing (compared to the forces of our body) is an indestructible essence, drying up all the superfluities of our bodies, and has been philosophically called by the above-mentioned name. It is neither hot and dry like fire, nor cold and moist like water, nor warm and moist like air, nor dry and cold like earth. But it is a skillful, perfect equation of all the Elements, a right commingling of natural forces, a most particular union of spiritual virtues, an indissoluble uniting of body and soul. It is the purest and noblest substance of an indestructible body, which cannot be destroyed nor harmed by the Elements, and is produced by Art. With this, Aristotle prepared an apple, prolonging life by its scent, when he, fifteen days before his death, could neither eat nor drink on account of old age. This spiritual Essence, or One Thing, was revealed from above to Adam, and was greatly desired by the Holy Fathers; this also Hermes and Aristotle call the Truth without Lies, the most sure of all things certain, the Secret of all Secrets. It is the Last and the Highest Thing to be sought under the Heavens, a wondrous closing and finish of philosophical work, by which are discovered the dews of Heaven and the fastnesses of Earth. What the mouth of man cannot utter is all found in this spirit. As Morienus says: "He who has this has all things, and wants no other aid. For in it are all temporal happiness, bodily health, and earthly fortune. It is the spirit of the fifth substance, a Fount of all Joys (beneath the rays of the moon), the Supporter of Heaven and Earth, the Mover of Sea and Wind, the Outpourer of Rain, upholding the strength of all things, an excellent spirit above Heavenly and other spirits, giving Health, Joy, Peace, Love; driving away Hatred and Sorrow, bringing in Joy, expelling all Evil, quickly healing all Diseases, destroying Poverty and misery, leading to all good things, preventing all evil words and thoughts, giving man his heart's desire, bringing to the pious earthly honour and long life, but to the wicked who misuse it, Eternal Punishment". This is the Spirit of Truth, which the world cannot comprehend without the interposition of the Holy Ghost, or without the instruction of those who know it. The same is of a mysterious nature, wondrous strength, boundless power. The Saints, from the beginning of the world, have desired to behold its face. By Avicenna this Spirit is named the Soul of the

World. For, as the Soul moves all the limbs of the body, so also does this Spirit move all bodies. And as the Soul is in all the limbs of the Body, so also is this Spirit in all elementary created things. It is sought by many and found by few. It is beheld from afar and found near; for it exists in everything, in every place, and at all times. It has the powers of all creatures; its action is found in all elements, and the qualities of all things are therein, even in the highest perfection. By virtue of this essence did Adam and the Patriarchs preserve their health and live to an extreme age, some of them also flourishing in great riches. When the philosophers had discovered it, with great diligence and labour, they straightway concealed it under a strange tongue, and in parables, lest the same should become known to the unworthy, and the pearls be cast before swine. For if everyone knew it, all work and industry would cease; man would desire nothing but this one thing, people would live wickedly, and the world be ruined, seeing that they would provoke God by reason of their avarice and superfluity. For eye hath not seen, nor ear heard, nor hath the heart of man understood what Heaven hath naturally incorporated with this Spirit. Therefore have I briefly enumerated some of the qualities of this Spirit, to the Honour of God, that the pious may reverently praise Him in His gifts (which gift of God shall afterwards come to them), and I will herewith shew what powers and virtues it possesses in each thing, also its outward appearance, that it may be more readily recognized. In its first state, it appears as an impure earthly body, full of imperfections. It then has an earthly nature, healing all sickness and wounds in the bowels of man, producing good and consuming proud flesh, expelling all stenches, and healing generally, inwardly and outwardly. In its second nature, it appears as a watery body, somewhat more beautiful than before, because (although still having its corruptions) its Virtue is greater. It is much nearer the truth, and more effective in works. In this form it cures cold and hot fevers and is a specific against poisons, which it drives from heart and lungs, healing the same when injured or wounded, purifying the blood, and, taken three times a day, is of great comfort in all diseases. But in its third nature it appears as an aerial body, of an oily nature, almost freed from all imperfections, in which form it does many wondrous works, producing beauty and strength of body, and (a small quantity being taken in the food) preventing melancholy and heating of the gall, increasing the quantity of the blood and seed, so that frequent bleeding becomes necessary. It expands the blood vessels, cures withered limbs, restores strength to the sight, in growing persons removes what is superfluous and makes good defects in the limbs. In its fourth nature it appears in a fiery form (not quite freed from all imperfections, still somewhat watery and not dried enough), wherein it has many virtues, making the old young and reviving those at the point of death. For if to such an one there be given, in wine, a barleycorn's weight of this fire, so that it reach the stomach, it goes to his heart, renewing him at once, driving away all previous moisture and poison, and restoring the natural heat of the liver. Given in small doses to old people, it removes the diseases of age, giving the old young hearts and bodies. Hence it is called the Elixir of Life. In its fifth and last nature, it appears in a glorified and illuminated form, without defects, shining like gold and silver, wherein it possesses all previous powers and virtues in a higher and more wondrous degree. Here its natural works are taken for miracles. When applied to the roots of dead trees they revive, bringing forth leaves and fruit. A lamp, the oil of which is mingled with this spirit, continues to burn forever without diminution. It converts crystals into the most precious stones of all colours, equal to those from the mines, and does mayn other incredible wonders which may not be revealed to the unworthy. For it heals all dead and living bodies without other medicine. Here Christ is my witness that I lie not, for all heavenly influences are united and combined therein. This essence also reveals all treasures in earth and sea, converts all metallic bodies into gold, and there is nothing like unto it under Heaven. This spirit is the secret hidden from the beginning, yet granted by God to a few holy men for the revealing of these inches to His Glory - dwelling in fiery form in the air, and leading earth with itself to heaven, while from its body there flow whole rivers of living water. This spirit flies through the midst of the Heavens like a morning mist, leads its burning fire into the water, and has its shining realm in the heavens. And although these writings may be regarded as false by the reader, yet to the initiated they are true and possible, when the hidden sense is properly understood. For God is wonderful in His works, and His wisdom is without end. This spirit in its fiery form is

called a Sandaraca, in the aerial a Kybrick, in
the watery an Azoth, in the earthly Alcohoph and
Aliocosoph. Hence they are deceived by these
names who, seeking without instruction, think to
find this Spirit of Life in things foreign to
our Art. For although this spirit which we seek,
on account of its qualities, is called by these
names, yet the same is not in these bodies and
cannot be in them. For a refined spirit cannot
appear except in a body suitable to its nature.
And, by however many names it be called, let no
one imagine there be different spirits, for, say
what one will, there is but one spirit working
everywhere and in all things. That is the spirit
which, when rising, illumines the Heavens, when
setting incorporates the purity of Earth, and
when brooding has embraced the Waters. This
spirit is named Raphael, the Angel of God, the
subtlest and purest, whom the others all obey as
their King. This spiritual substance is neither
heavenly nor hellish, but an airy, pure, and
hearty body, midway between the highest and
lowest, without reason, but fruitful in works,
and the most select and beautiful of all other
heavenly things. This work of God is far too
deep for understanding, for it is the last,
greatest, and highest secret of Nature. It is
the Spirit of God, which in the Beginning filled

the earth and brooded over the waters, which the
world cannot grasp without the gracious
interposition of the Holy Spirit and instruction
from those who know it, which also the whole
world desires for its virtue, and which cannot
be prized enough. For it reaches to the planets,
raises the clouds, drives away mists, gives its
light to all things, turns everything into Sun
and Moon, bestows all health and abundance of
treasure, cleanses the leper, brightens the
eyes, banishes sorrow, heals the sick, reveals
all hidden treasures, and, generally, cures all
diseases. Through this spirit have the
philosophers invented the Seven Liberal Arts,
and thereby gained their riches. Through the
same Moses made the golden vessels in the Ark,
and King Solomon did many beautiful works to the
honour of God. Therewith Moses built the
Tabernacle, Noah the Ark, Solomon the Temple. By
this, Ezra restored the Law, and Miriam, Moses'
sister, was hospitable; Abraham, Isaac, and
Jacob, and other righteous men, have had
lifelong abundance and riches; and all the
saints possessing it have therewith praised God.
Therefore is its acquisition very hard, more
than that of gold and silver. For it is the best
of all things, because, of all things mortal
that man can desire in this world, nothing can

compare with it, and in it alone is truth. Hence
it is called the Stone and Spirit of Truth; in
its works is no vanity, its praise cannot be
sufficiently expressed. I am unable to speak
enough of its virtues, because its good
qualities and powers are beyond human thoughts,
unutterable by the tongue of man, and in it are
found the properties of all things. Yea, there
is nothing deeper in Nature. O unfathomable
abyss of God's Wisdom, which thus hath united
and comprised in the virtue and power of this
One Spirit the qualities of all existing bodies!
O unspeakable honour and boundless joy granted
to mortal man! For the destructible things of
Nature are restored by virtue of the said
Spirit. O mystery of mysteries, most secret of
all secret things, and healing and medicine of
all things! Thou last discovery in earthly
natures, last best gift to Patriarchs and Sages,
greatly desired by the whole world! Oh, what a
wondrous and laudable spirit is purity, in which
stand all joy, riches, fruitfulness of life, and
art of all arts, a power which to its initiates
grants all material joys! O desirable knowledge,
lovely above all things beneath the circle of
the Moon, by which Nature is strengthened, and
heart and limbs are renewed, blooming youth is
preserved, old age driven away, weakness

destroyed, beauty in its perfection preserved, and abundance ensured in all things pleasing to men! O thou spiritual substance, lovely above all things! O thou wondrous power, strengthening all the world! O thou invincible virtue, highest of all that is, although despised by the ignorant, yet held by the wise in great praise, honour, and glory, that - proceeding from humours - wakest the dead, expellest diseases, restorest the voice of the dying! O thou treasure of treasures, mystery of mysteries, called by Avicenna "an unspeakable substance", the purest and most perfect soul of the world, than which there is nothing more costly under Heaven, unfathomable in nature and power, wonderful in virtue and works, having no equal among creatures, possessing the virtues of all bodies under Heaven! For from it flow the water of life, the oil and honey of eternal healing, and thus hath it nourished them with honey and water from the rock. Therefore, saith Morienus: "He who hath it, the same also hath all things". Blessed art Thou, Lord God of our Fathers, in that Thou hast given the prophets this knowledge and understanding, that they have hidden these things (lest they should be discovered by the blind, and those drowned in worldly godlessness) by which the wise and the pious have praised

Thee! For the discoverers of the mystery of this thing to the unworthy are breakers of the seal of Heavenly Revelation, thereby offending God's Majesty, and bringing upon themselves many misfortunes and the punishments of God. Therefore, I beg all Christians, possessing this knowledge, to communicate the same to nobody, except it be to one living in Godliness, of well-proved virtue, and praising God, Who has given such a treasure to man. For many seek, but few find it. Hence the impure and those living in vice are unworthy of it. Therefore is this Art to be shewn to all God-fearing persons, because it cannot be bought with a price. I testify before God that I lie not, although it appear impossible to fools, that no one has hitherto explored Nature so deeply. The Almighty be praised for having created this Art and for revealing it to God-fearing men. Amen. And thus is fulfilled this precious and excellent work, called the revealing of the occult spirit, in which lie hidden the secrets and mysteries of the world. But this spirit is one genius, and divine, wonderful, and lordly power. For it embraces the whole world, and overcomes the Elements and the fifth Substance. To our Trismegistus Spagyrus, Jesus Christ, be praise and glory immortal. Amen.

The Treasure of Treasures

by

Philippus Theophrastus Bombast,

Paracelsus the Great

As also The Water-Stone of The Wise Men;
Describing the matter of, and manner how
to attain the universal Tincture. Faithfully
Englished. And Published by J.H. Oxon.
London, Printed for Giles Calvert,
and are to be sold at the Black Spred Eagle,
at the West end of Pauls,
1659.

The Treasure of Treasures

NATURE begets a mineral in the bowels of the
earth. There are two kinds of it, which are found
in many districts of Europe. The best which has
been offered to me, which also has been found
genuine in experimentation, is externally in the
figure of the greater world, and is in the eastern
part of the sphere of the Sun. The other, in the
Southern Star, is now in its first efflorescence.
The bowels of the earth thrust this forth through
its surface. It is found red in its first
coagulation, and in it lie hid all the flowers and
colours of the minerals. Much has been written
about it by the philosophers, for it is of a cold
and moist nature, and agrees with the element of
water.

So far as relates to the knowledge of it and
experiment with it, all the philosophers before me,
though they have aimed at it with their missiles,
have gone very wide of the mark. They believed that
Mercury and Sulphur were the mother of all metals,
never even dreaming of making mention meanwhile of a
third; and yet when the water is separated from it
by Spagyric Art the truth is plainly revealed,
though it was unknown to Galen or to Avicenna. But

if, for the sake of our excellent physicians, we had to describe only the name, the composition; the dissolution, and coagulation, as in the beginning of the world Nature proceeds with all growing things, a whole year would scarcely suffice me, and, in order to explain these things, not even the skins of numerous cows would be adequate.

Now, I assert that in this mineral are found three principles, which are Mercury, Sulphur, and the Mineral Water which has served to naturally coagulate it. Spagyric science is able to extract this last from its proper juice when it is not altogether matured, in the middle of the autumn, just like a pear from a tree. The tree potentially contains the pear. If the Celestial Stars and Nature agree, the tree first of all puts forth shoots in the month of March; then it thrusts out buds, and when these open the flower appears, and so on in due order until in autumn the pear grows ripe. So is it with the minerals. These are born, in like manner, in the bowels of the earth. Let the Alchemists who are seeking the Treasure of Treasures carefully note this. I will shew them the way, its beginning, its middle, and its end. In the following treatise I will describe the proper Water, the proper Sulphur, and the proper Balm thereof. By means of these three the resolution and composition are coagulated into one.

CONCERNING THE SULPHUR OF CINNABAR.

Take mineral Cinnabar and prepare it in the following manner. Cook it with rain water in a stone vessel for three hours. Then purify it carefully, and dissolve it in Aqua Regis, which is composed of equal parts of vitriol, nitre, and sal ammoniac. Another formula is vitriol, saltpetre, alum, and common salt.

Distil this in an alembic. Pour it on again, and separate carefully the pure from the impure thus. Let it putrefy for a month in horse-dung; then separate the elements in the following manner. If it puts forth its sign[20], commence the distillation by means of an alembic with a fire of the first degree. The water and the air will ascend; the fire and the earth will remain at the bottom. Afterwards join them again, and gradually treat with the ashes. So

---

[20] The Sign is nothing else than the mark left by an operation. The house constructed by the architect is the sign of his handicraft whereby his skill and art are determined. Thus the sign is the achievement itself. - De Colica.

the water and the air will again ascend first, and afterwards the element of fire, which expert artists recognize. The earth will remain in the bottom of the vessel. This collect there. It is what many seek after and few find.

This dead earth in the reverberatory you will prepare according to the rules of Art, and afterwards add fire of the first degree for five days and nights. When these have elapsed you must apply the second degree for the same number of days and nights, and proceed according to Art with the material enclosed. At length you will find a volatile salt, like a thin alkali, containing in itself the Astrum of fire and earth[21]. Mix this with

---

[21] The earth also has its Astrum, its course, its order, just as much as the Firmament, but peculiar to the element. So also there is an Astrum in the water, even as in the earth, and in like manner with air and fire. Consequently, the upper Astrum has the Astra of the elements for its medium and operates through them by an irresistible attraction. Through this operation of the superior and inferior Astra, all things are fecundated, and led on to their end. - Explicatio Totius Astronomiae. Without the Astra the elements cannot flourish. ... In the Astrum of the earth all the celestial operations thrive. The Astrum itself is hidden, the bodies are manifest. ... The motion of the earth is brought about by the Astrum of the earth. ... There are four Astra in man (corresponding to those of the four elements), for he is the lesser world. - De Caducis, Par. II.

the two elements that have been preserved, the water and the earth. Again place it on the ashes for eight days and eight nights, and you will find that which has been neglected by many Artists. Separate this according to your experience, and according to the rules of the Spagyric Art, and you will have a white earth, from which its colour has been extracted. Join the element of fire and salt to the alkalized earth. Digest in a pelican to extract the essence. Then a new earth will be deposited, which put aside.

CONCERNING THE RED LION.

Afterwards take the lion in the pelican which also is found [at] first, when you see its tincture, that is to say, the element of fire which stands above the water, the air, and the earth. Separate it from its deposit by trituration. Thus you will have the true aurum potabile[22]. Sweeten this with the alcohol of wine poured over it, and then distil in an

[22] Aurum Potabile, that is, Potable Gold, Oil of Gold, and Quintessence of Gold, are distinguished thus. Aurum Potabile is gold rendered potable by intermixture with other substances, and with liquids. Oil of Gold is an oil extracted from the precious metal without the addition of anything. The Quintessence of Gold is the redness of gold extracted therefrom and separated from the body of the metal. - De Membris Contractis, Tract II., c. 2.

alembic until you perceive no acidity to remain in the Aqua Regia.

This Oil of the Sun, enclosed in a retort hermetically sealed, you must place for elevation that it may be exalted and doubled in its degree. Then put the vessel, still closely shut, in a cool place. Thus it will not be dissolved, but coagulated. Place it again for elevation and coagulation, and repeat this three times. Thus will be produced the Tincture of the Sun, perfect in its degree. Keep this in its own place.

CONCERNING THE GREEN LION.

Take the vitriol of Venus[23], carefully prepared according to the rules of Spagyric Art; and add thereto the elements of water and air which you have reserved. Resolve, and set to putrefy for a month according to instructions. When the putrefaction is

---

[23] If copper be pounded and resolved without a corrosive, you have Vitriol. From this may be prepared the quintessence, oil, and liquor thereof. - De Morbis Tartareis. Cuprine Vitriol is Vitriol cooked with Copper. - De Morbis Vermium, Par. 6. Chalcanthum is present in Venus, and Venus can by separation be reduced into Chalcanthum. - Chirurgia Magna. Pars. III., Lib. IV.

finished, you will behold the sign of the elements. Separate, and you will soon see two colours, namely, white and red. The red is above the white. The red tincture of the vitriol is so powerful that it reddens all white bodies, and whitens all red ones, which is wonderful.

Work upon this tincture by means of a retort, and you will perceive a blackness issue forth. Treat it again by means of the retort, repeating the operation until it comes out whitish. Go on, and do not despair of the work. Rectify until you find the true, clear Green Lion, which you will recognize by its great weight. You will see that it is heavy and large. This is the Tincture, transparent gold. You will see marvelous signs of this Green Lion, such as could be bought by no treasures of the Roman Leo. Happy he who has learnt how to find it and use it for a tincture!

This is the true and genuine Balsam[24], the Balsam of the Heavenly Stars, suffering no bodies to decay,

[24] There is, indeed, diffused through all things a Balsam created by God, without which putrefaction would immediately supervene. Thus in corpses which are anointed with Balsam we see that corruption is arrested and thus in the physical body we infer that there is a certain natural and congenital Balsam, in the absence of which the living and complete man would not be safe from putrefaction. Nothing removes the

nor allowing leprosy, gout, or dropsy to take root. It is given in a dose of one grain, if it has been fermented with Sulphur of Gold.

Ah, Charles the German, where is your treasure? Where are your philosophers? Where your doctors? Where are your decocters of woods, who at least purge and relax? Is your heaven reversed? Have your stars wandered out of their course, and are they straying in another orbit, away from the line of limitation, since your eyes are smitten with blindness, as by a carbuncle, and other things making a show of ornament, beauty, and pomp? If your artists only knew that their prince Galen - they call none like him - was sticking in hell, from whence he has sent letters to me, they would make the sign of the cross upon themselves with a fox's tail. In the same way your Avicenna sits in the vestibule of the infernal portal; and I have disputed with him about his aurum potabile, his Tincture of the Philosophers, his Quintessence, and Philosophers' Stone, his Mithridatic, his Theriac,

---

Balsam but death. But this kind differs from what is more commonly called Balsam, in that the one is conservative of the living, and the other of the dead. - Chirurgia Magna, Pt. II., Tract II., c 3. The confection of Balsam requires special knowledge of chemistry, and it was first discovered by the Alchemists. - Ibid., Pt. I., Tract II., c. 4.

and all the rest. O, you hypocrites, who despise the truths taught you by a true physician, who is himself instructed by Nature, and is a son of God himself! Come, then, and listen, impostors who prevail only by the authority of your high positions! After my death, my disciples will burst forth and drag you to the light, and shall expose your dirty drugs, wherewith up to this time you have compassed the death of princes, and the most invincible magnates of the Christian world. Woe for your necks in the Day of Judgment! I know that the monarchy will be mine. Mine, too, will be the honour and glory. Not that I praise myself: Nature praises me. Of her I am born; her I follow. She knows me, and I know her. The light which is in her I have beheld in her; outside, too, I have proved the same in the figure of the microcosm, and found it in that universe.

But I must proceed with my design in order to satisfy my disciples to the full extent of their wish. I willingly do this for them, if only skilled in the light of Nature and thoroughly practised in astral matters, they finally become adepts in philosophy, which enables them to know the nature of every kind of water.

Take, then, of this liquid of the minerals which I have described, four parts by weight; of the Earth of red Sol two parts; of Sulphur of Sol one part. Put these together into a pelican, congelate, and dissolve them three times. Thus you will have the Tincture of the Alchemists. We have not here described its weight: but this is given in the book on Transmutations[25].

So, now, he who has one to a thousand ounces of the Astrum Solis shall also tinge his own body of Sol.

If you have the Astrum of Mercury, in the same manner, you will tinge the whole body of common Mercury. If you have the Astrum of Venus you will, in like manner, tinge the whole body of Venus, and change it into the best metal. These facts have all been proved. The same must also be understood as to the Astra of the other planets, as Saturn, Jupiter, Mars, Luna, and the rest. For tinctures are also prepared from these: concerning which we now make no mention in this place, because we have already dwelt

----

[25] It is difficult to identify the treatise to which reference is made here. It does not seem to be the seventh book concerning The Nature of Things, nor the ensuing tract on Cements. The general question of natural and artificial weight is discussed in the Aurora of the Philosophers. No detached work on Transmutations has come down to us.

at sufficient length upon them in the book on the Nature of Things and in the Archidoxies. So, too, the first entity of metals and terrestrial minerals have been made, sufficiently clear for Alchemists to enable them to get the Alchemists' Tincture.

This work, the Tincture of the Alchemists, need not be one of nine months; but quickly, and without any delay, you may go on by the Spaygric Art of the Alchemists, and, in the space of forty days, you can fix this alchemical substance, exalt it, putrefy it, ferment it, coagulate it into a stone, and produce the Alchemical Phoenix[26]. But it should be noted well that the Sulphur of Cinnabar becomes the Flying Eagle, whose wings fly away without wind, and carry the body of the phoenix to the nest of the parent, where it is nourished by the element of fire, and the young ones dig out its eyes: from whence there emerges a whiteness, divided in its sphere, into a sphere and life out of its own heart, by the balsam of its inward parts, according to the property of the cabalists.

HERE ENDS THE TREASURE OF TREASURES.

---

[26] Know that the Phoenix is the soul of the Iliaster (that is, the first chaos of the matter of all things). ... It is also the Iliastic soul in man. - Liber Azoth, S. V., Practica Lineae Vitae.

The Manual of the Stone of the Philosophers

By THEOPHRASTUS of Germany,
Called PARACELSUS the Great.

If you would (by VULCAN'S Art) frame the PHILOSOPHERS STONE, which for very weighty causes we call a perpetual Balsom, you are first of all to know and consider how that STONE is to be placed materially before thy Eyes, and be made visible and sensible; and likewise how the virtue or fire thereof may come forth and be known. But for the clearer setting forth of this my intention, let's borrow an Example of Common Fire, viz., By what means the virtue or power thereof shews itself, and becomes visible: And that is done on this wise: First of all by the Vulcanick Art is Fire smitten out of the Flint: Now indeed this Fire can do nothing unless it hath as is friendly to it, upon which 'tis capable to operate, such as is Wood, Rosin, Oyle, or such like things, as it is naturally easily combustible; When therefore the said Fire lights upon some such a like thing, it proceeds on to an unceasing operation, unless the fuel fail by which it multiplies itself; for if Wood or such like be put thereto, then doth its force become stronger and shews its operation in the Wood, even so is the same thing done with the STONE OF THE PHILOSOPHERS, or the perpetual Balsom in Man's Body.

IF that STONE be made by a prudent Phisician, out of a convenient matter, and by a philosophick manner, and (after a due consideration of all the circumstances of man) it be administered unto him, it doth then renovate and restore the Organs of Life, in such wise as if the Wood were put to a Fire, by which the well-nigh deed Fire is cherished, and a shining and clear Flame procured: From hence therefore 'tis evident that there's much placed in the matter of this Balsom, forasmuch as 'tis behoveful that it have a singular Harmony with the body of man, and may so exercise its virtue, as that the Human Body may be safe from all the accidents as might be able to happen thereunto from such a matter.

AND therefore there is not onely much placed in the preparation of the STONE or Balsom, but 'tis much more behooveful to know the true matter itself, that is thereto fit; and furthermore, to prepare and use it as is fitting, viz., soberly and prudently, that so such a Medic may be able to purge away all the defilements of the Blood, and other superfluities, and may induce Health in the room of the Disease. 'Tis therefore expedient for a true and honest Phisician, to have a good knowledge, and not to regard ambition and pomp, not to appoint things doubtful or contrary, nor to trust an Apothecary too much, but well to know the Disease and the Diseased, or otherwise ye will always heal sinisterly, and will get nothing thence-from, but onely this, viz. The sick is deluded, and only deceived by the pride and ignorance of the foolish & unmet Physician. But this is a great sin, and such as will not go unpunished. For what is it else but a voluntary wickedness, viz. For anyone to demand money, and a reward for that which he knows Nothing of, and yet he'll be a Master, but with infamy enough; For many men do dis-esteem money, and not regard it, could they but be rightly advised and informed: But if this be not done, they lose both their Bodies, and Fortunes: And yet nevertheless 'tis counted a praise to demand money and fees; but believe it he that lists, for my part I'll reckon of such a Doctor after another-guess manner: For 'tis manifest, that of such Doctors (who in their own conceits are most highly learned) there's not a tenth part that hath a right knowledge of Simples, and much less that are certain of what they command to be done, and how the Medicine is boiled by the Apothecary; so it often comes to pass, that such a Doctor orders such or such a simple to be taken in his composition, which himself never knew, and haply the Apothecary much less; and verily it frequently is so, that the Apothecary hath it not at all; and yet this Medicine must be called Perfect, and the sick must drink it off as a good Medicine, and pay deer enough for it, but as to the Event, that the Patient feels' for although it be no way as profitable to him as his health, yet 'tis profitable to the Doctor and Apothecary, as to the filling of their Purses. But if the Doctor and Apothecary themselves should be possessed with the like Disease, they would not take such a Medicine: Therefore it may well be guest, how miserably and wickedly they act, and that 'tis most highly necessary for them to order their Affairs otherwise, to amend their errors, and to follow better things: But I fear 'twill be a hard matter to tame and master old Dogs.

BUT to return to my purpose, (from whence a just zeal to the miserable and forlorn sick persons withdrew me) and to give satisfaction concerning it, I say, that it is not so much expedient onely to prate or boast of the PHILOSOPHERS STONE, but 'tis necessary that the STONE be framed and prepared out of a convenient Matter, and be discreetly used: But know, that many of the Ancients have in their parabolical writings sufficiently discovered that Matter, and have, moreover, disclosed the Operation under figurative expressions, but yet have not wholly, and perfectly, manifested it; that so the foolish ones might not abuse it, and yet their sons might not have it hidden from them.

BUT whereas they are but a few that have followed them, and that have aptly set upon the thing, these secrets have, in process of time, been as 'twere, blotted out of remembrance, and Galenical fables have crept into the room thereof: But as the foundation thereof was at first laid, so doth it even yet stand in the same state, or rather grow daily worse and worse: This you may see in their Herbaries, how do they torment themselves therein? How do the GERMANES mix ITALY with GERMANY; whenas, notwithstanding, GERMANY doth not need those ultramarine herbs, but hath even sufficient of perfect medicine in herself. And therefore lest the truth should be constrained to give place to a Lye, and lest the darknesses of GALEN, and his Complices, should quench or suppress the Light of nature in medicine, it is expedient for me THEOPHRASTUS to speak in this little book, not as an imaginary Physician, but as a knowing one, and as such an one as is not ashamed of his actions in Medicine, and who (by the grace of God assisting) have had good proof and experience in many sick persons, such as thou Galenist durst not have visited: Tell me now, thou Galenical Doctor, from whence came thy foundation? Dost thou not place the bridle upon the horse's tail? Didst thou ever cure the Gout? Didst thou ever dare to go to the Leprous? Hast thou cured the Dropsie? I believe, and that upon good reason, that thou wilt be mute, and suffer THEOPHRASTUS to be thy master. But if thou wilt learn, learn and see what I shall here write and say; viz. That the body of man hath no need of thy Herby chariot, especially in Chronick and long continued diseases, the which (by reason of Ignorance) thou cellest wholly incurable; for thy Herbs are too weak for these diseases, and cannot, of their own nature, find out the Centre of the disease.

NEITHER wilt thou be able to do anything with thy Pills, unless to purge the Excrements onely; and withal, because of their inconveniency, thou oft expellest the good with the bad; the which cannot possibly be done, the great damage of the sick; and therefore well might those Pills ; Furthermore, neither do thy Syrups profit anything, yea rather are as a thing of no value; and bring such a nauseate to him that takes them, because of their horrible and loathsome savour, that they burthen the sick, and do afterwards induce gripings and danger, but do operate against nature.

BUT now I'll leave the rest of thy absurd and improper medicaments, for that they fight directly against nature, nor should be made use of by any means. Whereas therefore, those things that I have spoken are true, and that there's no true medicine to be found in Galen, Rhasis, or Mesue, that can set upon the said diseases in their root, and purge them out, (even as the fire mundifies the skin of the Salamander;) it necessarily follows, that the Cure of THEOPHRASTUS is far different from the Galenical Fantasies, and that his Curing flows forth from the Fountain of Nature; otherwise THEOPHRASTUS should be as reproach-worthy as they.

IF therefore we should follow Nature, and use natural medicine, let us see what things they are, amongst all others, that are most convenient in medicine for the body of man; viz. for the Conserving it (by reason of their virtue and efficacie) in soundness and health, even to the term of the predestinated death.

THIS then, if considered of, I doubt not but that 'twill be on all hands Confest, that Metalline things have the greatest agreement with the body of man, and that the perfect Metals, by reason of their perfection, but principally their radical humidity, are able to do much upon the body of man: for that a man is also a partaker of that SALT, SULPHUR, and MERCURY, which doth in some measure, though hidden, rest in metals, and metalline things. Now then like is to be applied to like, the which so wonderfully profitable to nature, so it be rightly done, the which is a great secret in Medicine, yea, may be called an ARCANUM.

WHAT wonder therefore, is it, if excellent, unheard of, and inseparable Cures do follow, and such as ignorant men accounted impossible to be done?

BUT that I may not digress any further, I must for brevities sake, here hint that I have determined to write in this Book; for I have a mind of treating more clearly here in this place concerning true medicine, then elsewhere. But we have afore told how man hath his original of SULPHUR, MERCURY and SALT, even as Metals have; this therefore being sufficiently declared in the book, PARAMIRUM, 'tis needless to repeat it here; wherefore I shall only shew you, how the aforesaid STONE OF THE PHILOSOPHERS may be known, and in some measure prepared: Know therefore of a Certainty, that there's nothing so small, out of which any thing is to be made, that can stand without Form; for all things are Formed, Generated, Multiplied, and destroyed in their Concordancy, and proper agreeableness, and do shew their Originality, whereby it may be perceived, what it was in the beginning; and that, that which runs, or steps in between, is like to that imperfection which nature admixeth in the Generation.

BUT whereas such Accidents may be separated by VULCAN, least they might do somewhat that may be an hindrance, Nature may in this case be amended, and this is likewise done in this Stone; for if thou wouldst make it of the right proper matter (the which may be well enough known by the aforementioned circumstances) 'tis necessary that thou take from it its superfluities, and frame, multiply, and augmentation in its Concordancy, or uniformity, like another, or third thing: for without its Condordancy it cannot be thus done, for Nature hath left it imperfect here, forasmuch as it hath not framed the Stone, but the proper Matter, and is hindered by accidents, whereby 'tis incapable of doing those things that the Stone, when prepared, is able to effect: and therefore such a Matter, without preparation, is, in respect of the STONE, but an half and imperfect thing, and stands not in any Concordancy, and Harmony, whereby it may be called perfect, or may be helpful for the health of man's body. The MICROCOSM affords thee an example of this thing: for behold, man as formed by the working Framer, into a man only, is not an whole & perfect-work, whilst standing out of his concordancy, but is but an half-work, until the framing of the woman suitable to him, and then he is a complete entire work.

BUT both of them are Earth; and so now these two Earths do constitute, or make up an entire man, capable of Augmentation and Growth; and this is done by the thus framed Concordancy. In such wise must be done with the PHILOSOPHERS STONE, if you would have it Renovate as well Men as Metals: If it be unburdened of its superfluous Accidents, and be placed in its Concordancy, it causeth wonderful effects in all Diseases: Except this be done, all your Attempts thereabouts are but vain.

BUT now if you would thus place it into its Concordance, 'tis expedient that you reduce it into its First Matter, that so the Male may act upon the Female, and that its outside may be turned inwards, and its inside outwards, and that both the seeds, viz., the Male and Female, may be enclosed in their own concordancy, and be by VULCAN'S help brought to a more than perfect condition, and be exalted in their degree, and withal, may from itself pour in all virtue, (as being a clarified, temperate, and qualified Essence) into man's Body, & likewise into Metals, and may render them sound & whole, driving out all the defilements by way of expulsion, and that the good in the blood of man may thereby be drawn to the due places, by the means of attraction, that SO the MICROCOSM which is posited in the LIMBUS of the Earth, and framed of the Earth, may by this Medicine as being like himself be Radically, and not imaginarily, but most assuredly be restored to health, or preserved therein. This therefore, is a Mystery of Nature, and such a secret as every Physician ought necessarily to know; And indeed every one that's born of the ASTRAL Medicine is capable of comprehending it; But that I may more clearly describe the Matter, and the preparation of a Medicine so excellent, that so an entrance may be given to the Sons of Learning, who love the truth.

YOU must know, that Nature hath given us a certain thing in which (as in a chest) are mysteriously concluded, or comprehended 1. 2. 3. The virtue and power whereof is more than enough sufficient for the conservation of the health of the MICROCOSM, insomuch that after preparation, it expels all imperfections, and is a true Defensive against old Age, and by us called a Balsom.

BUT now you must first know what thing it is that Nature hath placed such a number in: for I cannot describe it thee more clearly, for many reasons.

BUT as to the preparation thereof, neither GALEN, RHASIS, or MESUE, understood it, nor that those who follow them attain it: For this Medicine hath such a preparation, as your Pill-sellers attain not unto, and much less for an HELVETIAN-CALF to apprehend. Moreover it hath as it were celestial and singular operations; for it doth purifie and renovate by (as 'twere) a regenerating way as you may read more at large in my ARCHIDOXIS; and withal, well and advisedly take notice of the Original and the Essence, together with the virtue of Metals and

Metallick things. He therefore that hath ears to hear, let him hear and see whether or no he speaks groundlessly and from the Devil, as thou Sophister triflest and supposes, who art thyself invironed with the Devil, Lyes and Darkness, and callest nothing Good but what thy foolish head is able to comprehend, and what serves for thy fancy without any previous labour: For thou seest but with one Eye, and erroneously wandrest, nor goest thou to the right Window of the Kitchen: But yet thou maist without offending me, wind about thy intricate thread, and search for the Centre of the Labyrinth amongst the dark stars. But notwithstanding, if thou shalt at any time hap to make use of thy Wisdom, and consider what thing the PARACELSIAN-Art is founded upon and how lame thy hotch-potch-fragments are, there would not be that contrariety betwixt thee and PARACELSUS: For, as concerning the things whereof I now do, and shall briefly write, whereby my ASTRAL Disciples may apprehend and enjoy them, and glory of them; these things I say, may by the diligence of such as one as is not ashamed to learn, be well understood, there being nothing so difficult but may be known, and learned by labour and study.

THE practice therefore of this Work, is as follows:

THE PREPARATION OF THE MATTER OF THE STONE:

TAKE the Mineral ELECTRUM, filed; put it into
its own SPERM, (Others read it thus: TAKE THE
IMMATURE MINERAL ELECTRUM, PUT IT INTO ITS OWN
SPHEARE:) that the defilements and superfluities
thereof may be washed away, and purge it to the
utmost, as much as you can with STIBIUM, after a
Chymical manner, lest that otherwise thou shouldst
suffer loss, by reason of its impurity. Then resolve
it in the stomach of the Ostrich which is born in
the earth, and is comforted and strengthened in its
virtue, by the sharpness of the Eagle. But when the
ELECTRUM is consumed, and hath, after its solution,
gotten a Marigold-colour, be not unmindful of
reducing it into a spiritual transparent essence,
resembling the colour of true Amber: then add half
so much onely of the extended Eagle as the corporal
ELECTRUM (afore its preparation) weighed, and oft
times abstract thence-from the stomach of the
Ostrich, for so thy ELECTRUM will be still more and
more spiritual. But when the stomach of the Ostrich
is wearied, or spent with labour, 'tis needful that
thou refreshest, or renewest it, and from time to
time abstractest it. Then lastly, when it hath again
lost its sharpness, add the TARTARIZATED
Quintessence, yet in such a proportion, as to overtop it the height of four fingers, that so it may be
deprived of its redness, and may pass, or distill

142

over together therewithal: this do so long and so often, until it becomes white of itself: Now then when 'tis enough, (for thou wilt see with thine eyes, how it will by little and little fit itself for sublimation) and thou perceives that sign, sublime it, and so the ELECTRUM will be converted into the whiteness of the exalted Eagle, and 'tis brought thus to pass, and is transmuted by a little labour. This now is that we seek for, for our use in Medicine; with the which thou mayest make a safe progress in many Diseases, which will not yield to vulgar medicines. Thou mayest likewise convert this same into a water, an oil, or a red powder, and make use thereof in all such medicinal cases as need requires.

Give me leave to tell thee; and that truly, that there is not a better foundation for the whole structure of Medicine, then what lies hid in the ELECTRUM. Albeit I do not deny that (according to what I write in my other Books) there lie hid even in other Mineral things great secrets, but then they require a longer and greater labour, and besides 'tis more difficult to use them aright, especially for the unskillful; for if such make use of them, there accrues more hurt than good thencefrom.

FOR these respects therefore, it is not laudable for every Alchymist to exercise the Medicinal Art, if he be ignorant thereof. It would be expedient that, as to this, some let and bar were here instituted, that so an inhibition might be imposed on such putatitious, imaginary Physicians. For my part, I'll not bear their blame, nor acknowledge them for disciples, seeing they follow not the truth; but account of them as notorious deceivers, and slotheful Loyterers, such as snatch the bread out of the true disciples' mouths, and of set purpose hurt men, esteaming neither Conscience nor Art. But in our said ELECTRUM prepared, there lies do great a virtue of Curing men, that there cannot be found a more certain and more excellent medicine in the whole world.

INDEED the Galenical Triacle-selling Doctors do call it poison, and oppose it, not knowingly, but out of pride and mere foolishness. I myself do likewise grant that in its preparation it is a poison, and as great, or greater than that of the Tyrian Serpent, or Adder, that is put into Triacle; but that it remains poisonous after its preparation, that is as yet indemonstrable; for though to sole blockish brains it be incomprehensible, yet doth nature always tend unto its perfection, and it may therefore be much rather brought to that pass by convenient Arts, then alone. But I'll grant yet farther, that after its preparation, it is a greater venome, and more vehement then afore; but yet 'tis onely such a venome as is so directed, as to seek after its like, and to find out fixed and other incurable diseases, and expell them; not in such wise, as to suffer the Disease to be operative, and so hurtful, but 'tis, as it were, an enemy to the disease, and attracts to itself the like matter, and radically absumes it; and it doth so wash, even as Soap scoureth off the spots in foul cloathes, and together with the said spots, doth itself also go off, and leaves the cloathes purified, unhurt, clean, and fair.

SO then, this venome (as thou callest it) hath a far other and better efficacie than thy AXUNGIA, which thou art wont to make use of, in the cure of the French disease, and which thou oftner anointest withal, then the Currier doth his skins. For this ARCANUM which lyes hid in this Medicament, hath in its self a well-proportioned, well-prepared, and excellent essence; such as admits not of any Comparison with other poison, unless you apprehendme according as myself said before; and it is as much different from thy ARGENT VIVE, which thou anointest with; and from thy Precipitate, as to virtue and efficacy, as the heaven is from the earth. 'Tis therefore called, and indeed is a medicine blessed by God, and is not revealed unto all; for 'tis much better corrected then those mucky, dirty medicaments that the slow-paced Doctor hath in his gown, or hath filtred through his double Streiners, or Fools bag:

Furthermore, this blessed Medicine hath thrice greater force and operative virtue in all diseases, whatsoever name they are called by, they have all the Store-houses and Shops thou ever sawest. But now I attained not hereunto by idleness, sitting still, and sloth, nor did I find it in an Urinal; but by Travelling, and (as thou termest it) Wandering: I perceived that if I would indeed know, and not conjecture onely, 'twas necessary for me to learn by much diligence and labour. But as for thee, thou suckest thy Medicine and Art out of the old Mattress, or Pallet, old Cushion, or Cuch, wherein the Necromentick Old Wife sitteth; 'tis shee who hath inspired thee, and hath covered thy Coelestial Intellect with a BLEW CAP for Medicine: It doth not therefore at all repent me of my Journeyings, for I shall continue to be thy Master, and trace the steps of MACHAON, which spring forth from the LIGHT OF NATURE, even as a flower doth by the heat of the Sun. But that the Work I have intended may not be retarded, and be left imperfect, we shall go on to observe how the procedure is to be made, and what virtue and property Medicinal Nature hath given to this PHILOSOPHICAL STONE, and how it may be brought to the end.

THE RESIDUE OF THE PREPARATION FOLLOWS:

THY ELECTRUM being destroyed, as aforesaid, if thou wouldst make a farther progress, and arrive to thy wished end, Take the destroyed and flying-made, or Volatized ELECTRUM, as much of it as thou hast a desire to perfect, and put it in a Philosophical Egg, and seal it excellently well, that nothing may evaporate: Let it stand so long in an ATHANOR, until it doth of itself, without any addition, begin to be resolved from above; in such wise that there be an appearance of, as 'twere, an Island in the midst of that sea, the which doth day by day grow less and less, till at last it be changed into the blackness of Shoemakers'-black, or Ink: This black is the Bird which flies without wings by night, the which even the first celestial dew hath by a perpetual Coction, and ascension, and descension, transmuted into the blackness of the head of a Crow, the which doth afterwards assume the Peacock's tail, and then gets the feathers of the Swan, and last of all, receiveth the highest Redness of the whole world; the which is a sign of its fiery nature, by the which fire it expels all the Accidents of the body, and cherisheth the cold and dead embers.

SUCH a Preparation as this is done (according to the saying of all Philosophers) in one onely

VESSEL, one FURNACE, one FIRE, the VAPOROUS FIRE never ceasing.

SO then, that Medicine is as 'twere Celestial and Perfect, or at least may be made a more than perfect ___ or MEDICINE, by its own proper Flesh and Blood, and by its internal Fire, produced and turned outwards, as was spoken of but now, whereby both all the defilements of Metals are washed away, and by which also the hidden parts of Metals are manifested: For that same More-than-perfect Medicine can do all things, it penetrates all things, and infuseth, or pours in health in that very self-same time when it expels the Evil, and Disease. Therefore there's no Medicine in the Earth that is like it. Herein then exercise thyself, and be strong, for this is it which will bring thee praise and glory; neither wilt thou be an imaginary, but a knowing Physician; yea, thou wilt be even constrained to love thy Neighbor; for such a Divine ARCANUM cannot be perceived or understood by anyone without Divine Assistance, not its virtue, for 'tis unspeakable and infinite, in, and by which the Omnipotent God is to be known.

BUT know, that there's no Solution made in thy ELECTRUM, unless it hath perfectly run through the Circle of the seven Spheres thrice; for this Number becomes it, and this Number it must fulfil: Give heed therefore to the Preparation, for 'tis the cause of Solution, and to the Glorified, destroyed, and spiritualized ELECTRUM, use the TARTARIZATED ARCANUM to wash off the superfluities which happened in the Preparation, lest you labour in vain: But yet notwithstanding, nothing of the ARCANUM of TARTAR will remain there, but you are to proceed with it onely circularly, according to the aforesaid Number; for so it easily becomes of itself in the Philosophical Egg, and Vapour of the Fire, a Philosophical Water, the which the Philosophers call a Viscous-Water: It will also coagulate itself, and represent itself in all colours, and at last be adorned with the highest Redness.

I AM prohibited to write more plain of this Mysterie, it is at the Dispose of the Divine power; for this Art is most assuredly the Gift of God, and therefore all men cannot understand it, God bestows it on whom it pleaseth Him, nor will He suffer it to be forcibly wrested out of His hand, but will alone have the Honour herein: Whose Name be blessed for evermore: AMEN.

NOW FOLLOWS THE USE OF THE STONE

'TIS likewise expedient that I write of the USE of this Medicine, and its Weight: Know therefore, that the DOSE of this Medicine is so little and small, that it is scarcely credible, and that it must be taken only in Wine, or the like; but however taken, it must be of the very smallest Quantity, because of its heavenly force, virtue, and efficacy; for it is onely for this end manifested unto man, that so no imperfection may remain in Nature; and it is so provided and predestinated by God, that the Virtue and ARCANUM thereof may be produced by Art, to the intent that all creatures may be constrained to be profitable unto man, as being God's Image; but above all, that the Omnipotency of God may be made known. He therefore that receiveth his understanding from God, to him shall this MEDICINE be given: But the ignorant GALENICAL Drone will never be able to comprehend it, but rather loath and abhor it; for all his Works are Darknesses, whereas this Work doth operate and act in the Light of Nature. Thus in few, but true words, hast thou the Root of all true Medicine, and its Original, such as nobody shall pluck from me; no, though RHASIS with all his foul off-spring be staring-mad; though GALEN be as bitter as Gall, and AVICEN gnasheth his teeth, and MESUE lies largely, yet it will be too high for them all, and THEOPHRASTUS will stand firm in the truth: Whereas on the other side, the maimed works of the APOTHECARIES, and the smearings of the PHYSICIANS, together with all their pomp and foundation, will tumble down.

ONE thing more 'tis convenient for me to speak, because my present Writing will seem obscure to many: thou wilt haply say, MY Theophrastus, THOU SPEAKEST TOO BRIEFLY AND INTRICATELY UNTO ME, I KNOW NOT THY KIND OF SPEAKING, AND HOW RIGHTLY THOU DECLAREST THY THINGS AND ARCANAES, THIS WRITING OF THINE WILL NOT PROFIT ME AT ALL. Hereto I answer thus: Pearls belong not to Swine, nor a long tale to a Goat, for Nature would not give it them; therefore I say, He to whom God will vouchsafe it, he shall find sufficiently, and more than enough, yea more then he hath been desirous of.

I WRITE these things for an entrance and beginning, follow thou on prudently, neither shun thou study, labour, or the Coales: Let not the bragging praters seduce or hinder thee, nor turn thee aside from that diligence which is requisite, for by perpetual MEDITATIONS, are many fruitful and profitable things found out: Wherefore accept of what I give thee in good part, and apply thyself to make use of the Fountain, so shalt thou have no need to drink out of the pits of the Philosophers, nor shalt thou have anything to do with the dead-buryers, but shalt be able to serve thy Neighbour well, and bring praise and honour to God: He that is a master of Hares-dung, even so let him remain, with him is neither help nor counsel. These things I was willing briefly to set down in this my little Book of the PHILOSOPHERS STONE, lest men might imagine that THEOPHRASTUS cures many Diseases by Diabolical means: If thou followest me aright, thou shalt do the same, and thy MEDICINE shall be like unto the Ayre, which pierceth thru all open things, and is in all things driving forth all six Diseases, and inmixing itself Radically, whereby health may succeed in the place of the Disease: For out of this Fountain doth TRUE AURUM POTABILE abundantly flow, nor can better be anywhere found.

ACCEPT of these Instructions as a faithful Admonition, and do not reject and vilifie THEOPHRASTUS before thou knowest what he is: I am unwilling to set down anything else in this Book, though it would have been necessary to say somewhat of AURUM POTABILE, and to speak of the Liquor of SOL, philosophically: But I shall onely hint at these things, the which if they are but rightly prepared, are not to be contemned in their virtues: But because my other Bookes do treat much, and declare sufficiently enough as to these things, viz. What a true Physician ought to know, I will even here forbear, hoping that this little Book will not be altogether fruitless, but rather be a sufficient Counsellor to the Sons of Learning.

THE LORD BESTOW HIS GRACE FOR HIS OWN GLORY AND HONOUR: AMEN.

PARACELSUS his way of Extracting MERCURY Out of METALS.

TO extract a MERCURY out of Metalline Bodies, is nothing else but to resolve them, or reduce them into their first Matter, that is, into RUNNING-MERCURY, viz., such a MERCURY as it was in the CENTRE of the EARTH before the GENERATION of METALS, viz., a moist and viscous VAPOUR, containing in it the MERCURY and SULPHUR of Nature invisibly, which are the principles of all the metals, such a MERCURY is of unspeakable Virtues, and possesseth Divine Secrets.

THIS Reduction is made by a MERCURIAL WATER, which neither JOHN de RUPESCISSA, nor others, whatever boasts they have made thereof, ever knew: It is therefore by much diligence to be known, and to be handled or improved with unwearied Labour. On this wise therefore is the said MERCURIAL WATER to be prepared.

TAKE MERCURY SEVEN TIMES SUBLIMED, WITH VITRIOL, SALT-NITRE AND ALLUM, THREE POUNDS.

OF SAL ARMONIACK CLEAR AND WHITE, THRICE SUBLIMED FROM SALT, ONE POUND AND A HALF.

Both of them being ground together, and ALCOLIZATED, sublime them in a SUBLIMATORY in SAND for nine hours: Let all cool; then take off the SUBLIMATE with a Feather, and sublime it with the remainder, as before: Repeat this ono more, and the MASSE remain black in the bottom, and flowing like to wax: cool it, and take it out, and grind it again, and put it in a Glass-Dish, and imbibe it oftentimes with the water of SAL ARMONIACK, but let it be the prepared Water, and let it Coagulate of its own accord, and then again imbibe it, and dry it, even until nine or ten times, until it will be almost no more coagulated. Then grind it subtilly upon a Marble, dissolve it in a moist place, into a fair Oyle, the which you must rectifie by distillation in Ashes, from all fece and residence. Diligently keep this most chief and principal Water, of the which--- TAKE eight Ounces, and put thereinto one Ounce and a half of most purely mundified Leaves or Plates of the best SOL or LUNA, set it in a digesting in hot ashes, in a shut Glass, for eight hours: Thou shalt see thy body in the bottom of the Vessel, transmuted into a subtile VAPOUR, or MERCURY: The Solution being made of the whole Mercurial water, separate it from the first Matter by Sublimation in an Alembick, with a gentle fire, and keep it in a glass vessel diligently. By this means shalt thou have the most true Mercury of a body, the use whereof in desperate Cases (provided it be wisely used) is miraculous, and celestial; and therefore not to be revealed to the unworthy.

THE SULPHUR OF METALLS, by Theophrastus:

The SULPHUR OF METALLS is an Oilyness extracted out of them, the which is endued with abundance of virtues for the health of Man. One Sulphur is extracted out of Metalls afore they have endured the fire: as for example, Out of golden, or silver MARCASITES, &c., according to the nobility of the Mineral, so is the Sulphur noble and excellent. So likewise out of the Mineral of a MARCASITE, and COBALT, each as its nature and propriety is.

The more common way of the extraction thereof is this: viz. You must take Vinegar excellently well distilled, such as hath stood for 24 hours upon a caput Mortuum of Vitriol, Salt Nitre, and Allum distilled, and been distilled by an Alembick thencefrom. This, I say, must thou pour upon the pulverized metalline body, in a glass that it may overtop it seven fingers, and set it to digest in a Horse-belly for nine days: then take the coloured Vinegar, and distill it in ashes, even to the Sulphurous oil, the which you must rectifie in a B.

or at the ⚗ , so shalt thou have a most true Sulphur of the metalline body, the which thou must rightly make use of according to thy discretion. An extraction may likewise be made by a sharp and well-depurated LIXIVIUM. But these other Sulphurs are not so profitable, as to the inward use of the body, because of the ALKALI of the Ashes, out of which we make a clavellated gnawing Salt: and because of the CALX, or Limyness which such LIXIVIUMS are made withal. The Sulphur that is thus extracted, may be washed with sweet water, and be Precipitated: but the digestion afterwards requires twice the time.

The LIXIVIUM also is to be rectified by the sublimation thereof from all terrestrial residence left that such Sulphurs being incorporated with it become corrosive, to the destruction of the sick: the which to prevent, the said separation is to be made. And so much concerning the Crude Sulphurs.

But now as for the melted and depurated Metalls, you may extract their Sulphur too: There is not given a more certain, a more noble, and better way; then by the water of Salt, or its Oile prepared on such wise as I have evidently described in my books of Alchemy: for such a Water doth fundamentally and radically extract out of all Metalline bodies, their natural liquor, or Sulphur, and most excellent CROCUS, as well for Medicinal, as for Chymical Operations: it resolveth and breaketh every metal, bringing it out of its Metalline nature into another, according to the different intention and industry of the workman.

The CROCUS, or Tincture of Mettalls:

The CROCUS of Metalls is four-fold: viz., the CROCUS of SOL, VENUS, MARS and CHALYBS: that of CHALYBS, or Steel, is the better, 'tis extracted by Reverberation, or Calcination, reducing the said bodies into a powder. So Iron filed is consumed by Rust: the Consumption of the Rust is made by the inhibition of those things which cause Rust, and by a dedoction extracting the colour of the Rust.

Take old Urine powred off from its residence, viz., some Cups of it, in which dissolve three handfuls of Salt ground; strain it, boil it, and scum it well: Herein again dissolve an handful of Vitriol beaten, and two Ounces or three of SAL ARMONIACK beaten, and scum it again very well: With this Liquor imbibe the filings, and boil them until they are pulverisable; the which powder you must reverberate with a strong fire, continually stirring it with an Iron rod, until it pass from one colour to another, and at last into a most lucid Violet colour; out of which thou mayest easily extract the Tincture by spirit of Wine, or distilled Vinegar, and by the separation of the Elements gather the abstraction left in the bottom of the vessel, with which thou mayest effect wonderful works, as well without the body, as without.

As for the making of CROCUS VENERIS, do thus: TAKE ONE POUND OR TWO OF THE BEST COPPER-RUST, or VERDIGREASE ALCOLIZATED; POUR THERETO PLENTY OF DISTILLED VINEGAR AND STIR IT WELL THRICE EVERY DAY. POUR OFF THE COLOUR'D VINEGAR GENTLY, AND TOTALLY SUBLIME, OR DISTILL IT IN ASHES, EVEN TO A DRYNESS: Let this powder be afterwards WASHED NINE TIMES WITH WARM WATER FROM ALL THE SHARPNESS AND BE DRIED; so shalt thou have the prepared CROCUS of VENUS, or the FLOS of BRASS; out of which thou mayest, if thou wilt, easily extract an Oil, according to the precepts delivered in the great Chirirgical work, where also its use is explained.

The CROCUS of ☉ is to be extracted by the water of Salt, by which the Metalline nature thereof or Malleation, is broken, the residence is to be washed with hot water, and the CROCUS to be extracted with spirit of Wine, the which being again separated, the CROCUS will remain in the bottom; the which by elevation, by the degrees of fire, in five times sublimation, is changed into a Liquor, or the most true Quintessence of SOL. With this thou mayest perform miraculous things: but hereto is required not a putatitious, but an industrious and skillful Artist.

FINIS

OF THE TINCTURE OF THE PHILOSOPHERS

By Philippus Theophrastus Paracelsus

Chapter I

I, Philip Theophrastus Bombast doth say, that after, and according to the manifestation of Divine Grace, there are various ways found out for the attaining to the TINCTURE of the Philosophers, all which do finally belong and lead to the same scope and end; for HERMES TRISMEGIST, the Egyptian, set about this Work according to his own peculiar way. ORUS the Grecian observed the same Process. HALI the Arabian continued in his peculiar Method: But ALBERT MAGNUS, a German, followed a very tedious process: Every one of them proceeded according to their proper instinct and manner; but yet nevertheless they all arrived to One and the same End, viz., to a Long Life, so much desired by the Philosophers, and to an honest Sustenation and Conservation thereof in this Valley of Miseries. Therefore I, Theoph. Paracelsus Bombast, the Monarch of ARCANUMS, am (in this season) gifted by God with peculiar Endowments and that for this end, viz. That every Searcher after this high natural work, may have occasion and encouragement of imitating and following me, be he what he will, either Italian, Polonien, Frenchman, German etc., Come hither after me, all ye Philosophers, Astronomers, and Spagyrists, what high

Name soever ye be of, I will shew and open unto you, ye Alchemists and Doctors, exalted by me with most exceeding labours, that Corporeal Regeneration; I will teach you the TINCTURE, the ARCANUM, or QUINTESSENCE, in which the very Fundamentals of all Mysteries and Operations do lye hid; for verily a man may, and Ought to believe another, Onely in such things as he hath made trial of by the fire: If anyone shall introduce anything besides this kind of Experience into the SPAGYRICAL or MEDICINAL Art, there's no cause of giving any credit thereunto; For Experience testifies that by the Fire is made a separation of the true from the false; For verily, the Light of Nature is on this wise created, that by it may be made evident the PROBA, or trial of everything; but yet to such Only as walk in that Light. Now from that Light we will teach by most excellent demonstrations and shew that all such as have ever, before my time, entered upon this so difficult a PROVINCE merely with their peculiar Fancies, and acute Speculations, have to their own loss made trial of their foolishness; from this foundation of mine, therefore, many Rusticks have been made Nobles, and contrariwise, many Nobles have been put by their putatitious and opiniative Art, turned into Rusticks, such as have carried Golden Mountains in their Hands, before they have set their hands to the Coales. In the first place therefore, is to be learned, Digestions, Distillations, Sublimations, Reverberations, Extractions, Solutions, Coagulations, Fermentations, Fixations; and every Instrument requisite to this work, is to be known by use; such as are Glasses, Curcirbites, Circulatories, Vessels of HERMES, Earthen Vessels, BALNEAS, Wind-Furnaces, Reverberatories, and other such like. Also a Marble, Coales and Tongs. Thus at length mayest thou profit in ALCHEMY and MEDICINE: But as long as thy Fancy and Opinion adheres to thy Ficticus Books. Thou art not fit, or predestined to any of these things.

CHAPTER II
Of the Definition of the Subject and Matter
Of the TINCTURE of the PHILOSOPHERS.

Before I come to the Process of the TINCTURE, 'tis expedient that I lay open unto thee the Subject thereof; for this hath always been particularly hidden hitherto by the lovers of TRUTH. The matter therefore, of the Tincture, (see that you understand so according to a SPAGYRICAL sense) is a certain thing which doth by the Art of VULCAN, pass out of three, into One Essence, or may remain. But, that I may mention it by its proper Name, according to the use of the Ancients, 'tis by many called, the RED LION, but is known but by a few; this LION may by the help of Nature, and the Art of an Artist, be transmuted into a White Eagle; so that of ONE are made TWO, and so Lustrous, that the splendor og Gold shines not so bright to a SPAGYRIST, as this doth; those two have a brighter shine IF kept in ONE. If now thou dost not understand the use of the CABALISTS, and of the Ancient ASTRONOMERS, or their custom, thou art not at all ordained by God for the SPAGYRICK Art, nor art chosen by NATURE for the work of VULCAN, nor created to open thy mouth concerning the CHEMICAL ART: The matter therefore of the TINCTURE, is a most excellent PEARL, and a most precious TREASURE, and the most noble thing (next to the Manifestation of the Most High, and the Consideration of Mankind) that can be in the Earth; this is the LILY of ALCHIMY and of MEDICINE, which the PHILOSOPHERS have so accurately sought after; but because of the defect of the entire knowledge thereof, and its perfect Preparation, they arrived not to its perfect end; there is given us an Initiation onely of the TINCTURE, by their searchings and Experiences; but as for the true Foundation, which my COLLEAGUES are to imitate, is left to me, so that no body may commit their Disguises with our Intentions. Deservedly therefore do I (after my long experiences) correct and regulate the SPAGYRISTS, and separate the false and erroneous from the true; for I have by my long Inquiries and diligent Search, fund out such ways, by which I may justly reprove and change many things: But yet notwithstanding, had I found the Experiments of the Ancients to have been more excellent than mine, I would not at all have taken such great Labours as I have willingly undergone for the sake, benefit and honour of all honest ALCHIMISTS, &c. Having therefore sufficiently declared the subject of the TINCTURE in such wise as can scarce be done more faithfully between two Brethren; or indeed is lawful to be done more, I will proceed to its preparation; and having first set down the Experiences of the First Age, I will

also add my Inventions, to which the Age of Grace
and Mercy will at last adhere, whatsoever
Patriarchel Exemplars thou, O Sophisters, makest and
producest in the meantime, as the standards of thy
Philosophy.

Chap. III

Of the Process of the Ancients in order to the PHILOSOPHICK TINCTURE, and of a briefer Invention by P A R A C E L S U S

The Ancient SPAGYRISTS putrefied LILY for a PHILOSOPHICAL MONTH, and did afterwards distil thencefrom the moist spirits, until at length the dry spirits were elevated; the CAPUT MORTUUM they again imbued, and moistened with the mist spirits, and abstracted them often thencefrom by distillation, and that so long until the dry spirits were wholly elevated; then afterwards they united the abstracted Humidities, and the dry spirits together by a PELLICANE, three or four times, until all the LILY remained dry in the bottom.

Although that the first Experience gave this proceeding before fixation, yet nevertheless our Ancestors have thereby oftentimes perfectly obtained their desire; but yet they would have lighted on a shorter way of attaining to the treasure of the RED LION, had they but learned the Harmony of ASTRONOMY, with ALCHIMY, as I have demonstrated it in the APOCALYPS of HERMES. But whereas every day, as Christ speaks for the comfort of the faithful, hath a care proper to itself; the Labour of the SPAGYRISTS before my time, was grevious and very great; but now in this last Age, by the help of the inflowing of the Holy Spirit, 'twill be eased by my THEORY and PRACTICK, and will be declared to all those that shall constantly persevere in their workings with patience: For, I have tried the properties of Nature, its Essences and Conditions, and have known its conjunction, as well as its Resolution; and this is the highest and greatest thing in a Philosopher, never as yet made known to Sophisters.

When therefore the first Age gave forth the first Experience of the TINCTURE, the SPAGYRISTS out of one simple THING made two; but when that Invention did perish afterwards by a diligent and thorough search light upon the two Names of that simple THING, and styled it by One Word, viz. LILY, as being the subject of the tincture.

Then the Imitators of Nature putrefied this Matter for its time, even as the seed in the Earth is: For nothing can be born thereof, nor can any ARCANUM break forth, or be revealed before this corruption or putrefaction. Then afterwards they abstracted the moist spirits from the Matter until at length by the violence of the fire, the dry were likewise sublimed, that so by this way they might attain unto maturity, (like as the Countryman expects in the season of the year where one thing is wont to ascend after another, and so to fall away.) Last of all, Even as after the SPRING, the SUMMER comes, SO they incorporated those fruits and dry spirits, & brought the Magistery of the TINCTURE to that pass that it became ripe for the harvest, and disposed itself to Maturation.

Chap. IV.

Of the process concerning the TINCTURE OF THE PHILOSOPHERS, abbreviated by PARACELSUS.

The ancient SPAGYRISTS would not have needed such a prolix labour and tedious reiteration, had they learned their Work out of my School, and so attempted it; they would fully as well have obtained their desired End, with far less costs and labours: But now in this Season in which THEOPHRASTUS PARACELSUS is become the Monarch of ARCANUM'S, the time is now at hand of the invention of that which was hidden to all the SPAGYRISTS, that were before me. And therefore I say, take ONLY the blood of a Rosie colour of the LION, and the Glue of the EAGLE, the which after thou hast conjoined them together, coagulate them according to the old process, and thou shalt have the TINCTURE OF THE PHILOSOPHERS, which an infinite number have sought after, and but a very, very few have found.

Thou SOPHISTER, Will thou, nor will thou, this is a MAGISTRY in Nature itself, and a MAGNALE or wonderful thing of God above Nature, and a most precious treasure in this valley of miseries. If thou beholdest it extrinsically, it seems to be somewhat a vile thing to transmute another thing into a much more noble Body than it was before; But thou must even brook it, and confess that this is a

Miracle produced by a SPAGYRIST, who by the Art of his Preparation corrupts a visible externally vile body, out of which he excites another most noble and most precious Essence. If now thou hast likewise learned anything from the Aristotelian Light, or of us, or anything of SERAPIO'S Rules, come hither, and bring it forth (by experience) unto light, and preserve the Right of the Schools, as becomes a Lover of Honour, and a Doctor: But if thou knowest nothing, and canst do nothing, Why dost thou despise me as if I were an irrational HELVETIAN-CALFE, and callest me a WANDERING-VAGABOND? ART is a second NATURE, and a peculiar world, as Experience witnesseth, and demonstrates against thee and thy Idols: And therefore sometimes the ALCHIMIST compoundeth some simples, the which he afterwards corrupts according as his necessity requires, and thence prepares another thing; For so oftentimes, of many things is made One thing, the which is more efficacious, and doth more than Nature by herself is able to do, as is evidently apparent in GASTAYNUM, where **?**[27] is made of **?**; also in CARINTHIA, where **?** is made of **?**, and in HUNGARY **?** is made of **?**: I shall forbear to speak of other transmutations of Natural Things; they are well known to the MAGI, and

---

[27] Unfortunately, items marked **?** were blank spaces in the mimeographed document. I believe they are missing symbols. -pnw

brought to light, and are more wonderful than those things that OVID declares in his METAMORPHOSES. But that you may rightly understand me, you must seek your LION in the East, and your EAGLE in the South, for this our assumed or chosen Work: Thou wilt not find better Instruments than what Hungaria and HISTRIA do produce: But if thou desirest to bring it from Unity, by Duality, into Trinity; with an equal permutation and change of each, then you must direct your journey to the South, for so in CYPRUS shall you obtain your whole desire; concerning which we must forbear the making of any larger Discourse than what we have here at present declared. There are many more of those ARCANUMS as exhibit transmutations, although but a few know them, the which thought manifested by the Lord God to any one, yet the reporting of this Art doth not therefore presently break out, but the Omnipotent God doth together with it also give understanding, of concealing these and other such Arts until the coming of HELIAS the Artist, in which time there shall be nothing so occult, that it shall be revealed. Yet also visibly perceive (though indeed I have no reason to speak a word of these things, because some may deride it) that in the fire of SULPHUR is a great Tincture for Gems, the which doth exalt them to a more noble degree than Nature of herself could do: But as for that Gradation of

Metals and Gems: I shall omit the Discourse of them in this place, for I have abundantly enough written thereof in the SECRET OF SECRETS, and in the Book of the VEXATIONS OF THE ALCHEMISTS and in other places. And now, as I have begun the process of our Ancestors concerning the Tincture of the Naturalists, we'll perfectly conclude and finish it.

Chap. V.

Of the Conclusion of the Process of the Ancients,
Made by PARACELSUS.

Lastly, the ancient SPAGYRISTS did by a certain orderly augmentation of the FIRE, so long fix the Pellicanated and dried LILY, until it came back from blackness (with a permutation of passing through all the colours) to be as red as blood, and did therewith assume the property of the SAlAMANDER. Indeed they rightly proceeded in such a labour; and 'tis very fitting and expedient that everyone that aims at the getting of this Pearl, should proceed after the same manner. 'Twill be too hard a task for me to declare this more clearly unto thee, unless thou hast learned in the School of the Alchimists, to observe the degree of the fire, and also to change, or alter thy vessels: for then at length shalt thou see, that presently upon the heating of thy LILY in the Physical Egg, it will with wonderful apparitions be made blacker than the Crow: then afterwards, in success of time, 'twill be whiter than the Swan: & then lastly 'twill pass through a yellow colour, and become more red than any blood. Seek, seek, saith the chiefest SPAGYRIST, and you shall find; knock and it shall be opened unto you: It will be impious and unseemly to thrust meat into the mouth of a perfidious bird; let her rather fly hereto; even as myself, together with such as were before me, have been constrained to do: Follow the true Art therefore, for this will guide thee unto the perfect knowledge thereof: 'Tis not fit to make a larger or clearer addition of anything, then what I have afore spoken. Let thy Pharisaical Schools teach thee what they will from their unstable and slippery foundation; it reacheth not its end or scope.

Now at length when thou has been as accurately taught as possibly can be done by the Alchimical industry, then at length, will there be nothing in the nature of things so difficult, which may not be made manifest unto thee by the help of this Art: But verily nature, barely of herself, never brings unto light anything that is exalted to the height of its perfection, as we may here see in this place, from (and concerning) the unity or union of our duality. But a man must (by Spagyrical preparations) bring it to thyat pass, to which it was destined by nature. Thus much therefore let suffice concerning the process of the Ancients, and my correcting of the TINCTURE OF THE PHILOSOPHERS, as to what pertains unto its operation.

Furthermore, when we have that Egyptian, or Hermetical treasure in our hands, 'tis expedient that we convert it to our benefit; and this may be done after a twofold manner, by the Spagyrical

Magistery. The first way is the application of it to the Renewing of the body; the latter is the using it for the Transmutation of metals. And whereas I, Theophrastus Paracelsus, have diversely experienced them both, I am willing to describe and set the same down according to the signs of the work, and according as they have better and more perfectly appeared in, and by the testimony of Experience.

Chapter VI

Of the TRANSMUTATION OF METALS by the PROJECTION of the MEDICINE.

If you would make use of the TINCTURE OF THE PHILOSOPHERS for transmutation, then first of all, there must be one pound thereof projected upon a thousand pound of molten SOL; then at length will thy medicine be prepared for the transmutation of the leprous humidity, or juice of the metals. This is a wonderful work in the Light of Nature, viz. that by the Magistery, or Operation of the Spagyrist, a metal should perish from what it was afore, and become another. And this hath even rendered that same ARISTOTLE, together with his ill-founded Philosophy, a foolish man: for verily the Rustical fellows in HUNGARIA, when they have thrown some Iron into a certain Fountain, called, SIPFERDRONNEN and there left it, its time, 'tis consumed into a Rust, which being melted with a blast at the fire, it presently becomes pure VENUS, and nevermore returns into Iron. Likewise in the mountain RUTTENBERN commonly so called, they strain a LIXIVIUM, or a LY, out of MARCHASITES, in the which Iron is turn into most excellent, highly graduated VENUS, and more malleable than the other natural VENUS is.

These and many more such like things are better known to plain simple men, than to the Sophisters, viz. the transmutation of one SPECIES, and kind of metal into another. But yet these Arts, partly by reason of the very much contempt of the Ignorant, and partly because of the just envie and displeasure of Artificers, are almost quite hidden. Verily I have in ISTRIA, oftentimes brought VENUS beyond 24 degrees (ALIAS 58) so that the colour of SOL could not ascend higher, 'twas constant in the trial by Antimony and the Quartation, the which VENUS I have made use of, in all respects as the other.

But now although the Artists of old were very desirous of this ARCANUM, and sought after it with the highest diligence, yet nevertheless 'twas but a very few that could bring it by a perfect preparation to its end. For the transmutation of a meaner metal into a better, brings with it many difficulties and hindrances, as that of JUPITER into LUNA, or of VENUS into SOL. Haply 'tis the pleasure of God, that the MAGNALIA of Nature be hidden from many men, because of their sins; for when this TINCTURE hath been sometimes prepared by Artists, and that they could not bring their projection to take effect, it happened that by reason of their negligence and ill-keeping it, it was devoured by Hens, whose feathers thereupon fell off, and grew up again, this, I myself saw. By this way, through the abuse perpetrated by the negligence of the Artists, came Transmutation into Medicine and Alchemy; for, when they could not in the least make use of that TINCTURE according to their desire, they converted it to the RENOVATION of men, as you shall hear more at large in the following chapter.

Chapter VII
Of the RENOVATION of MEN.

Some of the chief and ancient Philosophers in Egypt lived by this TINCTURE an hundred and fifty years; likewise the lives of many men have been produced and prolonged for some Ages, as is most evidently mentioned in many histories; this will hardly seem credible to anybody, for the virtue thereof is so admirable, that it continues, and lengthens out the body more than is possible for its co-born nature to do; and it conserves it in that degree so firmly, that it lives safe and free from all infirmaties. And though it may have old age, yet nevertheless doth it appear as if constituted in its former juvenility.

The TINCTURE, therefore, of the PHILOSOPHERS, is an universal medicine, and consumes all diseases, whatsoever name they are called by, like an invisible fire: its DOSE is very little, but its operation is most powerful; I have thereby cured the Leapry, Lues Veneree, Dropsie, Falling Sickness, Collick, Gutta, and such like diseases: also the Woolf, Cancer, Noli-me-tangere, Fistula, and such kind of internal diseases, and that more certainly than is credible: concerning which, Germany, France, Italy, Poland, Bohemia, &c, will yield sufficient testimony. And now Sophister look back upon

Theophrastus Paracelsus; How can thy APOLLO, MACHAON, and HYPOCRATES be able to stand against me? This is the Catholicon of the Philosophers, by which all the Philosophers pursued long life, and resisted diseases; and did it by this universal TINCTURE most excellently, and most effectually obtain it, and styled it (according as seemed to them good) <u>THE TINCTURE OF THE PHILOSOPHERS</u> for what can there be greater in all medicine, than the mundifying of such a body, by which mundification all superfluity is even radically and totally taken away therefrom, and transmuted; for heal but the seed, and all things are become perfect. What profit is there in the most untowardly-founded purgation of the Sophisters, when as they take away nothing of that which they ought to remove. And therefore this is the most excellent foundation of a true Physician; viz. A Regenerating of nature, and a Restoring of youth; then afterwards the new essence itself, expels all that which is contrary to itself. In order to this Regeneration, the powers and virtue of the TINCTURE OF THE PHILOSOPHERS, are found to be wonderfully excellent, and are used with much secresie, and absconded by the true Spagyrists even to this time.

FINIS

Notes by HEB:

1. In the word SIPFERDRONNEN, the letter after the "D" is sufficiently indistinct as to suggest transposed letters might be implied. RONNEN might conceivably suggest "racing" with the PFERD-, which is Horse; on the other hand, the "S", may be meant for the indistinct letter, and we would have the Horse-Sun, or Horse of the Sun, or PEGASUS, especially as the Fountain is named.

2. In the case of the other "German" word, RUTTENBERN, this may be conceivably the Red of Amber, with a possible implication of the "rod of chastisement", or, in fine, the whip.

A BOOK OF
RENOVATION AND RESTAURATION

by THEOPHRASTUS, a Philosopher and Physician of Germany, called PARACELSUS the Great.

We are (in the first place) to understand, in the Creation of things, what RESTAURATION and RENOVATION are; what those things be, which restore and renovate, and also what this is that can be renewed and restored: Indeed all Minerals are thus brought unto a youthfulness, are renewed, and repaired, insomuch that rusty Iron may be again reduced into new Iron; and the Verdigreece, or flos Aeris, into its Copper: Likewise, Minium into Lead, and Saturn into Mercury (alias, the calx of Jupiter into Tin.) So then RENOVATION, and RESTAURATION (in this place) is that which reduceth a destroyed, or rusty, or consumed (Mineral) to its juvenility and perfect essence. But yet notwithstanding, this RENOVATION which we have here induced, cannot be compared (in the least) to that RESTAURATION, and RENOVATION which we pretend to expound: for although that Rust and Verdigreece be not a Metal, yet nevertheless, 'tis not as yet perished or consumed in its Metallic essence, and therefore it cannot (in this place) be made use of by way of comparison, for the explaining of our intent and meaning concerning RESTAURATION and RENOVATION, because such Rusts, nor Ablutions, doth not at all happen in Mankind; thence comes it to pass, that men do not at all need such a kind of Reduction.

But now if after this manner, a decreasing or consuming hap to befall an old, or decrepit man, (as if it were a kind of rustiness in his substance) then may his body be on such wise reduced from his decrepit degree unto juvenility, and it is a reduction from any whatsoever disease unto health; but yet this is not the thing which we (at present) will write of.[28] Moreover this also may be accounted of as a kind of Restoration, when a metal is naturally made out of Salt, Sulphur, and Mercury: This perfection being accomplished, and brought into a perfect metal, this metal may again easily return into its three first Principles, so that its Salt, Sulphur, and Mercury may again be made apparent, as they were at the first Generation thereof, insomuch that the metalline Essence may wholly pass away, and it be no more a metal. Thus may it also happen, viz. that the matter of the three Principles may return into a metal, as afore, as for example, if of the three Principles of Copper there be again made Copper, &c. This (now) in metals, is likewise a RESTAURATION or RENOVATION, when there is a certain

---

[28] More on this appears at the very end of this book. -pnw

generation made of a metal formerly perfect, into a perfect and complete metal again. But yet that is not to be esteemed of as a RENOVATION and restoration, if it be compared to a man; because we cannot (in the least) be reduced into our three Principles, or be brought into our SPERM out of which we may be again renewed and restored, (as we spoke above concerning metals.) For so it would come to pass, that we could then have power of bettering ourselves by a second generation, better than the first was, or as iron, which being reduced into its three first Principles, and afterwards into silver, or gold, is by this same made incorruptible, or as

♄ which is again reduced unto its own MERCURY, and at length changed into an incorruptible metal; Even so then should we be able to effect or create an incorruptible Creature out of ourselves, the which (notwithstanding) we have no power to do; for we want that first matter, nor can we go back into the irreducibility of our appointed and ordained state, but must proceed on, as we have begun, for there is no way by which we may be able to recover, or to have this thing out of which we proceeded.

There is therefore a twofold RESTAURATION or RENOVATION. One is what we have brought and declared about metals; The other is when an old Image is renewed with fresh colours, that so it appears fresh and new as it was at first; but yet we are not for this cause to understand in this place, that there is made a new matter out of the old; but that the old image is palleated, and cloaked as it were, so as to appear new; and therefore, neither may this be called a RESTAURATION, in reference to the RENOVATION and RESTAURATION of a man: But RENOVATION and RESTAURATION are to be understood after this manner, viz. That his Radical moisture acting or governing, and exercising the Spirit of life, be not diminished or driven backward, but be rather augmented and promoted in its virtues, as a tree that hath help administered unto it for the Production of its flowers and fruits, the which falling off, and being gone, there are others produced again as afore: But although that this example here made use of doth not on every side serve for the Declaration of our alleged opinion; yet nevertheless it affords us the understanding of making an advance or promotion of the Radical moisture of life, as we have demonstrated in the tree. And verily we would have you to apprehend of RENOVATION and RESTAURATION after this manner, viz. that it is not to be done in the radical humidity, but in that which is born of the said humidity, and draweth, or deriveth, its Originality, materially and corporally. For, even as a Bell made by fusion, doth not at all receive its sound from the TONE or note, that 'tis framed unto, but from the body; so the RESTAURATION or RENOVATION doth not receive its operation in the spirit of life, but in that which makes, and effects it; that is, the one is Material, and the other is substantial, or spiritual. But, when all this, in which the radical moisture is, shall be mundified, its TONE will be also mundified, and by how much the better its TONE is, so much the better will the body be. And when we say that the radical humidity proceeds from bodies and members, we intend and mean after this manner, That the radical moisture itself, and that which proceeds thencefrom, are; even as the root and the tree is; one whereof cannot at all live and subsist without the other: In like manner is it to be (here) understood, that these two are so united and conjoined, that they cannot be separated: therefore the radical humidity, and spirit of life, with the moisture of life, is in the bodies and members, even as the TONE or NOTE in a Metal, which is not seen, but only heard: for the spirit of life, and the radical humidity, are truly in bodies; therefore it will be a foolish thing for us to endeavor to amend it, or to renew the body by it, but 'tis expedient and fit, that the body and the matter, (which are born and risen, or sprung off, and with it) be renewed and restored. From thence, then, may it be collected, that RESTAURATION and RENOVATION is a Transmutation of those members, or parts, as exist

in the body, superfluously; so that every such thing
as proceedeth from the body and from the radical
humidity, may fall off, and new may be born in the
place thereof; even as we spake of <u>trees</u>, all the
<u>leaves</u> of which, its <u>flowers</u>, <u>fruits</u>, and
<u>excrescenses</u> do fall off, and again spring up, and
yet the <u>Wood</u> itself is not changed, so as to fall
off, and other to be born again, but remains; even
so likewise the <u>radical</u> humidity remains; that is,
the life in the body, and (with it) the body do cast
off the <u>hairs</u>, <u>nails</u> and <u>teeth</u>, and then afterwards
the like of them are <u>re-born</u>, and then grow again.
This therefore is <u>restoration</u>, and <u>renovation</u>, by
which this same thing as is to be renewed and
restored, is <u>so</u> <u>restored</u> and <u>renovated</u>; for every
RESTAURATION and RENOVATION is made in the
superfluities, and in those things that rise and
grow out of the substance: So then, it may be
sufficiently understood what way it is that the body
may be restored and renewed by: and (from the
demonstrations we have made) viz. from those
superfluities that are not of so <u>material</u> a growth,
as the hair, teeth, skin, and nails; for those are
in the body as certain superfluous things, and do
not pass into the corporal matters, or substance,
but remain in their own Essence, like as are the
FOUR COMPLEXIONS, alias humors, <u>one</u> whereof proceeds
from coldness and moisture, which is born and

retained in the whole body, and hath no peculiar place, nor beginning from whence it may proceed, as is proved (in our Discourse) of the FOUR COMPLEXIONS. A SECOND proceeds from things contrary to the former, viz. from heat and dryness, and even this also is <u>so</u> in the body, and hath no peculiar abiding or original, and it likewise causeth, or maketh moisture. The THIRD is cold and dry, and its way of birth is the same with the Others. The FOURTH is hot and moist, and doth proceed even as doth the rest.

Here you are to observe that it happens that all those FOUR HUMOURS are not always in all bodies, but sometimes one of them only, sometimes two, sometimes three, and otherwhile four; This also is to be heeded, as concerning them, that they are consumed and expelled in the RENOVATION and RESTAURATION, for this reason, because the nature and life of man can very well persist and abide without them, and doth not at all need them, for they are only superfluities; like as the <u>feces</u> are in the Wine, or as the scum and froth flowing therefrom in the Vintage-time.

This likewise is to be believed concerning the FOUR COMPLEXIONS appearing in a man, that they are not to be renovated and restored because they spring not from any either greater or lesser member: nor are they in the blood, nor in the flesh, or such

like: Neither is it true, that the sanguine
Complexion proceeds from the Liver, by reason of the
very much-abounding quantity of blood; or that
Melancholy proceeds from the Spleen, or Choler from
the Gall, and Phlegm from the Brain, and such like;
for verily the aforesaid Members do not give any man
his Complexion, but the Complexions happen to a man
even in his Nativity, and abide even till his death.
But we undertake not to dispute hereof in this
place, for as much as it would be <u>too</u> wide from our
text of RENOVATION and RESTAURATION.

Whereas therefore none of the Four Complexions
hath any place or original in the said bodies, but
they exist in the spirit of life, and in the <u>radical</u>
humidity: therefore the Complexions cannot be
renovated, nor restored; but where the body shall be
clarified, the nature of them is also clarified.

Moreover we also signify this, by our text,
viz. That the division, and distinction of
Complexions according to Age, Country, and Regiment,
is not at all to be cured, because there are no
Complexions imprinted in the body, from these three:
It may happen indeed that old Age may inforce a
sadness in bodies, but yet that is no Complexion: So
likewise the Country or Region may induce Phlegm,
but yet that Complexion is not therefore
Phlegmatick. So Choler may cause one to have a
yellow colour, but these things have no place of

discussing here, for they are declared in our treating of the <u>Construction</u>, or framing, <u>of the body</u>. There's a peculiar understanding requisite for Divisions (or Distinctions) of this sort, wherein is to be noted that they are not only Humours, but also sometimes Minerals, sometimes Corruptions; all which are superfluities against nature and strength: The like may be said concerning the principal Members, for these resist RENOVATION and RESTAURATION, thus, viz. these perceive not those (qualities of renewing, &c) nor receive they them unto themselves; but everything that passeth through them, and is prepared with, or by them; they receive and admit of it as Nutriment only, and not as a Medicament: but yet, if haply any humours or superfluities should be in them, they would be expelled. Even so then is it to be equally understood of the other Members, viz. the bones, marrow, brain, heart, liver, lungs, reins, spleen, stomach, intestines, gristles, muscle; and also as concerning the Blood, you are to know, that corruption, or superfluity, may be even in it; yet this is only accidental: And even so may it be equally-alike understood as concerning the Flesh; and verily this accident is purged away in the RENOVATION and RESTAURATION; not that there is to be made other blood, but that the evil be removed therefrom, and the good be preserved, and predominates. The same consideration is to be had concerning the flesh. But that we may briefly explain what those things are that may be restored and renewed, know that the Leprosie, Falling-sickness, Madness, Pustules, or Pox, the Gouts of the Feet, of the Hands, and of the Joints, and many other such like, may be taken away by the RENOVATION and RESTAURATION, unless there hap to be a Disease that had its Original even from the Nativity, for that will not be removed.

But as concerning the Leprosie, or if there be in the body any more grievous Disease than that is, you are to know that there is to be made a separation of the Pure from the Impure, but that the Leprosie be converted into Health, even as Copper and Iron are into Gold: Nor ought anyone to admire at this same TRANSMUTATION for RENOVATION and RESTAURATION do consume veen as the fire consumeth all the Falsities and Impurities that are in Gold or Silver, and leaves them pure and clean: By the same way are the Falling sickness & Gout removed; for so all things that are in the whole body are renewed by the flesh and blood, together with all the others therein concluded; for even as ALCALI mundifies the Leprous MERCURY into the best Silver, even so the RENOVATION and RESTAURATION do transmute the body into a good essence, as is said before.

So then RENOVATION and RESTAURATION drives forth whatsoever is superfluous in the body, and contrary to Nature, and changeth all that Nature doth not stand in need of, or which shall be of no moment or virtue, into good; Likewise it restores all things, and causeth them to grow again, as we said above: It reduceth the whole body into youthfulness, &c, and that for this reason, Because Nothing of those things as are in Nature itself, is able to resist them.

But now we come to consider the way by whiuch the body may be restored and renewed, viz. 'Tis done by that kindling (of a renewing and restoring Medicament) which it hath in the spirit of Life, and in the Radical humidity; by the which kindling the aforementioned Operations are made like to the burning virtues of a Nettle: Who is so quick-sighted as to be able rightly to search out such kind of virtues, when as they do not appear (in that action) so materially as they are sensibly known to be? After this manner also even RENOVATION and RESTAURATION of nature are as 'twere assisting-approaches made by such virtues as we are not able to express. Now we evidently know that every visible thing is cleansed and purged by fire; for so Nature requires that this very thing be done by fire, that is not possible to be done by any other thing. And therefore we understand a twofold fire, viz. A Material and an Essential Fire; the Material operates by a Flame, the Essential by the Essence and Virtues like CANTHARIDES that burn the skin, and raise blisters, like to the most violent fire: And yet notwithstanding they are not fire, (in the least) nor are they so perceptible to the sight, like as fire is: The same likewise doth Crowfoot and Nettles do, as we have oft times said.

'Tis in like manner evident unto us that the RENOVATION and RESTAURATION (when they come into the body, or are conjoined with it by union) do perfect their Operations after this way, viz. There is such an operation as is made in the MERCURY of SATURN or MARS, the which are put into the fire with their Realgars; and although neither of them be hot or fiery, yet they are burnt like wood, and the perfect Metal is found in the bottom, although it appeared altogether leprous before.

Likewise who is there that can search and find out what means it comes to pass by, that when MIGDALIO shall have been most vehemently lilted with VITRIOL, it becomes COPPER, and in all respects like to true COPPER, and yet it had not any similitude of COPPER afore: even so are we to understand concerning RENOVATION and RESTAURATION, viz. That they perfect their Operations like to Lime or Calx, which is extinguished or quenched with water, and purifies itself, and the force and acrimony thereof is taken away by the essential fire, and extinguished. The RENOVATION and RESTAURATION of our

Nature is much resembling that of the HALCYON, or KING-FISHER, the whence Bird is renovated by his own proper nature. Hence then, there are many more such-like things to be found as have a power of doing that, and of them we have made mention sundry ways in our ARCHIDOXIS, or much rather, in our SECRETS, from whence a very many might be brought, but that their digression from our present Text of RENOVATION and RESTAURATION, would be too much; such things as we there demonstrate, the same are to be understood in like manner here in this place concerning RENOVATION in our reiterated assertion, viz. That we cannot sufficiently or certainly know how the <u>fire operates</u>, although we see that it consumes the wood; for it overcomes and absumes all other things by the vehemency of its heat. But omitting this, we'll betake ourselves to another thing.

So then having abundantly enough spoken hitherto concerning the beginning of RENOVATION and RESTAURATION, let us now go on to discover those things which do not renovate and restore; We have indeed taught the preparation of them in our ARCHIDOXIS, and have given them their proper Names by which they may be known and heeded. Now we'll set down the Compositions of them, but in the first place their process; Now when we speak of, and teach you concerning simple Medicines and ARCANAES, 'tis to be understood that the operations thereof are done diversely; for there are some things to be found which do even violently cleanse the Leprosie, and do drive away no other Disease so well as they do that; and yet nevertheless are (as to RENOVATION and RESTAURATION) perfect; besides which, in the distinctions of Diseases of this kind, are the QUINTESSENCE, the MAGISTERY, and ELEMENT of ANTIMONY, the which doth so cleanse the body from the Leprosie, even as it doth purge Gold and Silver melted therein, in whom it leaves no footsteps of Impurity. So likewise the Element of SOL, and its Quintessence, as also its Oile, and AURUM POTABILE, do take away the Leprosie, together with all Diseases, and do renovate and restore; so likewise the Quintessence of Hellebor, of Celendine, of Bawm, Valerian, Saffron, Manna, and Betony, do renew the body, those Diseases abovementioned being excepted, for they do not drive them away.

Likewise the quintessence of Pearls, of Unio's, of the Smaragdine[29], the Saphire, Ruby, Granite, Jacynth, do renovate and restore the body into all perfection, they take away tartarous Diseases, and the Stone, Sand, Feet-Gout, Hand and Joint-Gout, and the things that are congealed and coagulated, and all such like Diseases as arise from TARTAR, so likewise the Quintessence and Magisteries of Minerals and of Liquors, do renovate and restore the

[29] Emerald -pnw

whole body without any defect, and free it from the Falling-sickness, Swounding, Suffocation, and all such Diseases as happen with a deprivation of the senses, as Madness, the Vitista, or Laughing Diseases, and such like.

The Magisteries and Essence of TARTAR, and of ALCALI, do also <u>renovate</u> the body with the perfection of RESTAURATION; they take away all Aposthumes, and amend the putrefactions and grossness of the Humors.

In like sort the Essences, Extractions, and Magisteries of the greater remedies, do <u>renovate</u> and restore the whole body; as for example, They remove Fevers, as Quotidian, Quartan, the Synochus, (or continual) the Ephemora Fever, &c. Likewise the first ENS of MARGARITES are able to <u>renew</u> and <u>restore</u> the whole body, and to take away all Womens Diseases, together with their Accidents, and to render both the Man and Woman fruitful; so likewise those same ARCANAS do take away all long and incurable Diseases by the <u>renewing</u> and <u>restoring</u> of the body into its supreme Virtues.

Thus also doth the Quintessence drawn out of BALSOM, <u>renovate</u> and <u>restore</u> the body, and take away Pleurisies and the Pestilence by the admirable operations and virtues of its perfecting property: There are many more such like things which shall be elsewhere repeated, and such as are of a far greater virtue than is able to be attributed unto them, and then can be mentioned. But yet as to these things, this is diligently to be heeded and considered of, as to Compositions, that although there are very many of them, yet none of them is sufficiently able generally to expel and cure all the Diseases (by itself) as are to be expelled by those Medicaments of RENOVATION. We'll therefore demonstrate the manner and Practick of our intention and meaning, even to the end; yet we will not set down all the Processes, for that is not necessary; for he that understands us, will apprehend and mark our writings, but he that doth not, will not be instructed by them; & yet we'll set down all the Processes (as are necessary) abundantly enough: But 'twould be burdensome for us to write down such things as many have written of, or are known before; yet this whole Doctrine cannot be better delivered, or treated of, than by the first ENS, wherein there's a singular Nature of operating upon the body, and of transmuting the essence thereof; for the first ENS itself is an imperfect Composition, predestinated to a certain and assured end, and corporal Matter. And because it is not perfect, it is able to alter everything that it shall be incorporated withal; even as MERCURY, which is like to the first imperfect ENS, as to imperfection: For although that it be determined and bounded, yet notwithstanding it is not changed from Imperfection, but is limited therewith.

MERCURY hath even a power of <u>renovating</u> the whole body, for that there is a most wonderful Laxative, & Alterative Virtue therein, which can never be sufficiently enough searched out: And yet nevertheless 'tis wholly imperfect and unprofitable in its OWN operation, and that (forsooth) because 'tis MERCURY, and the first ENS thereof must not be predestined into another's body; for such as it, viz. is, so is its perfection: But we speak of a first ENS that is perfect, for the <u>renovating</u> and <u>restoring</u> of the whole Body, as is the first ENS of Gold, nd that for this reason, because it truly possesseth the spirit of the Gols, and the most subtile spirit, and is far more subtile than the true body itself is, viz. GOLD.

Hence likewise the first ENS of SOL, or Gold is penetrable, even as a Mercury in Metals; nor containeth it in itself the Spirit of Salt, whereby it may be coagulated: for the Spirit of Salt coagulating the first ENS, doth take away so much virtue that the Gold is not a hundredth part so potent in its virtue, as its first ENS is; like to Generous Wine, which being coagulated with Frost, doth not again return to its former power and excellency.

But that we may perfectly speak and write concerning RENOVATION and RESTAURATION, you must know that the first ENS, that is, that the first composition of Gold, which is as yet a liquor, and not as yet coagulated, doth renovate & restore whatsoever it layes hold on; and not man only, but also all cattle, fruits, herbs, and trees: And what we thus mention, is to be understood on this wise, viz. Like to the Mineral, or Ore, of a Metal, which is endued with far greater virtues than its Metal is: for in the Ore there's a spirit of Arsenick, and Salt, of Sulphur and Mercury, all which do go away in the purging of the metal, the said metal remaining in one essence only.

The like is to be understood of the first Entities of Marcasites, as of Antimony; the which you must note to be no less than the first ENS of Gold is; for there is such a virtue in the first ENS of Antimony, that it doth, of itself, of its own proper nature, transmute everything that it seizeth on, even like as Antimony itself doth by the fire: for the virtue of it separates everything from the body that is born out of the radical humidity, and doth thoroughly renovate the said body from a true foundation; because its first ENS is seated in that predestination, insomuch that such as essence proceeds and goes out therefrom, as the heat doth out of the fire.

The like is to be understood of the first ENS of Rosins; The first ENS of Sulphur is an entire transmutation of the body into some RENOVATIONS and RESTAURATIONS, for it is so vehement that it tingeth all the first Entities of metals into their own essence, it takes away their operations, and reduceth them again into a new perfect body: For, the first ENS that's produced out of Sulphur, hath such virtues upon the body of a man, that it renews all the <u>radical</u> humidities in him, in all his parts and members.

In like sort may we speak of the first entities of Gems, the which do, by their first essence, most potently restore the whole body to its former powers and vigorousness, and do amend it of all its impurities, and renovate it, even as fire transmuteth Lead into a most pure Glass: for the <u>primum</u> <u>ENS</u> of the SMARAGDINE doth regenerate and renovate itself, for 'tis a perfect body from the beginning: even as the <u>green</u> <u>Marble</u>, the which hath such a nature from its proper predestination, that it renovates itself from all uncleanness and impurities, and doth again coagulate itself until it become pure; and it doth sometimes thus renovate itself, so much the more pure and constant doth it become: therefore as far forth as the virtues of the first Entities are known to me, they do assuredly far excel all the rest.

So likewise are you to note concerning the
first Entities of SALTS, that they are according to
their spiritual virtues, far greater than in their
perfection: So the first ENS of VITRIOL, transmutes
all white Metals whatsoever into red, and those that
are red into white, and overcomes and subdues all
the perfections that are comprehended in them; it
renovates and restores all the imperfect bodies of
the Metals, as TIN into its own first ENS, and into
TIN again, in which is more virtues than is in the
OLD TIN.

After the same manner it reduceth whatsoever
proceeds out of the radical humidities into the
radical moisture, and causeth the RENOVATION and
RESTAURATION itself more perfect, more plentiful,
and more abundant; for there's no other thing that
operates so vehemently upon the radical moisture.

Nor are the first Entities of herbs and trees
different from what was aforesaid, the which
Entities are a thousand times more potent as to
their virtues, than when they have received their
body, stalk, or trunk, and are incorporated: Even as
the first ENS of BAWM doth renovate and restore the
whole body far more powerfully than seems possible
to be done in natural things; for 'tis to be known
that the Halcyon, or King-fisher, is not thus
renovated nor restored from his own nature; but
because its nature is such as to be nourished and

live on first Entities, on this wise, when it feeds on the bodies of herbs, or seeds, and such like, his stomach doth, by digestion, reduce them to their first ENS, and doth afterwards out of that first ENS perfect the operations of its RENOVATION and RESTAURATION: for, that Bird's digestion hath its predestination naturally to first Entities only, whence it comes to pass that he doth first transmute all his food and drink into a first ENS; and therefore likewise doth he feed only on such bodies as to regenerate and restore, with which bodies he is even from the very beginning always provided for, and nourished with by his Parents, or Dam: this also is his nature, viz. to be renovated and restored after death; and that for this reason, because the first Entities cannot at all have their progress, or full course, in the Bird whilst he lives, for the life of this Bird takes away all the virtues thereof by converting them into blood and flesh; but being dead, he flourisheth according to the yearly seasons: And even as the first Entities disclose and produce themselves in the earth, even so, in like manner do they then, even in the Bird itself, put forth themselves and so renovate and restore the dead flesh; and this is (in nature herself) a very wonderful Argument of its most great virtues and power: And now, were not these things apparent to sight, they would seem incredible, although this described by many a one; for this cause also doth it happen that the HALCYONS do renovate themselves at different times, viz. some of them sooner, some later or slower, according as they have either more early or more late, eaten the first Entities; for some of them are born and do come forth either sooner or later than other-some do. In like sort there are very many Vermine or Worms renovated and restored, and that for this reason, Because they are as yet in the Earth, imperfect. Many more wonderful things are there that are hidden, yea far more than are known, or openly manifest, concerning which I could write more largely, but that it would be too much wide from the Text of the Book of RENOVATION and RESTAURATION.

And although we cannot so very well take, or get the first Entities, as we have written of them, or have them in the same Essence as we have demonstrated before, yet nevertheless 'tis a thing possible unto us; for if we know where the Mineral of Gold lies hid, we shall even there find its first ENS, if we but come afore its perfection; for there are certain signs whereby it may be known, in what manner the form of the Metal is posited, viz. thus: Whilst it is in its first ENS it makes trees fruitful, and the bottom, viz. the Earth, fertile; it renovates old trees, that have produced no fruits for these twenty years; for when the first ENS of

Gold shall lay hold on them, or on their Roots, they again begin to live and flourish as before; but albeit, that there are many more admirable things done by the first ENS of Gold, than we write of, yet notwithstanding these things are sufficient for the demonstration of the first ENS, that, viz. it is there.

But when you see flamings and Corruscations, or some sparklings, 'tis to be judged, and to be noted that the metal is made of the first ENS, and that it hath betaken itself into a Coagulation; these are to be accounted as signs, concerning the Original of the Minerals that appertain to Gold, Silver, or the other metals; for the signs of the first Entities of the other metals, as touching their original, are like those of Gold.

When therefore such a sign is seen, or found out, 'tis to be understood, that this very first ENS is not in the least so knit up (as 'twere) in one heap, as it is when it lies in its perfection, but is dilated (in that place) throughout that whole earth.

And therefore this earth is in the virtues of the first Entities, for out of it are they extracted: as 'tis in Celondine, when 'tis not as yet compounded, or fashioned: Its first ENS is in the earth, in which it hath its situation for this reason a like earth must be taken, and then it must be at length extracted therefrom, as we have demonstrated concerning the virtues of Celondine: you are likewise to observe, that there is this difference between the first ENS, and perfection: viz. that the first ENS can Renovate, and that for the reasons afore-mentioned: but being perfect, it hath only the natural virtues, so as to incline thitherward, as 'twere, but yet imperfectly. So then you are to understand from hence, that if you would have from them the virtues of those first Entities, then 'tis necessary that you remove them from their coagulation, and corrupt or break them, as is demonstrated in ARCANAS and Quintessences: but yet everything in its first ENS hath greater virtues. Nor let a Philosopher wonder hereat: viz. that out of a certain earth in which an herb is essentially born, before it be incorporated, all the virtues of that herb may be extracted: so that the virtues may be diligently kept, or preserved and the earth may again be put into its place, and in such wise, as that 'tis thenceforth but a mere earth, nor hath in it any fruitfulness at all, because its first ENS is now sequestered from it, that lay in the earth: from thence its wont to come to pass, that the virtue of such a first ENS, may be shut up in a glass, and be brought to that state, as that the form of that same herb may grow in itself without any earth; from whence it follows that the stalk thereof is nothing else but a certyain apparition to the sight, which may be again thrust down with your finger into a juice, in the likeness of a fume, the which demonstrates the Substantial form, but not perceptible by the touch. In such like growing things is the Quintessence altogether incorrupt, and in its highest perfection an in the earth.

Therefore there is born, after this manner, our of the first ENS of Gold, a concluded, or enclosed Gold, which in touch is like to a red water, and is stirred up, and is exalted after the manner of Gold.

But enough of this: Let's therefore now betake ourselves to the practick of those things as do Renovate and Restore; provided they be prepared according to the possibility and rule of Art: the which, though briefly described by us, yet are evident enough, for those intelligent men that have a good knowledge of Medicine and Philosophie.

So then, such things are to be known, in the first place, as Renovate and Restore, as we have demonstrated, and the first ENS of them is to be extracted, and by it is a Renovation and Restauration to be made: but for a close of this matter thus treated of, we'll set down four Mysteries: viz. of MINERALS, GEMS, HERBS and LIQUORS, as followeth.

The first ENS of MINERALS.

Take the mineral of Gold, or of Antimony, ground exceeding small, one pound; of circulated Salt, four pounds; mix them together, and digest them for a month in Horse-dung, then 'twill become a water, the pure whereof must be separated from the impure: coagulate this into a stone, the which you must calcine with cenesisted wine, and again separate it, and dissolve it upon a marble: putrefie this water for a month, then will there be made a liquor, in which do happen all the signs, as in the first ENS of Gold or Antimony, and therefore we justly call it the first ENS of these things: the same is to be understood concerning Mercury, and the others.

The first ENS of GEMS.

Take of SMARAGDI, or Emeralds excellently well ground, one dram, calcine them in Salt dissolved until they be converted into a whiteness; then let them be dissolved and be put into a Phial-glass, shut with the Lute of HERMES; let them be set over a naked fire, and let the matter be hang'd up somewhat high, in a bare uncoated glass, so as not to touch the bottom; and that so long, until it fall down from its spiritual nature and condition unto the bottom, into a body like the liquor of Honey. This body exhibits the virtues of the Emerald, and therefore may deservedly be called, the first ENS of the Emerald.

The first ENS of HERBS.

Take Celondine or Bawm; beat them into a Pultz, or mash, and put them in a glass shut with the Lute of HERMES; set them digesting in Horse-dung for a month, then afterwards separate the pure from the impure; pour out the pure into a glass with the dissolved Salt; the which being shut, let it be set in the Sun for a month, which time being over, thou shalt find a thick liquor in the bottom, and the salt swimming at top: separate it, and thou shalt have the virtues of the Bawm, or Celondine, as they are in their first ENS; the which both are. And are called, the first ENTITIES of BAWM or CELONDINE.

The first ENS of LIQUORS.

Take the Mineral of Sulphur, and Salt dissolved, and let them be totally reduced into a Water by themselves, the which distil four times: there will ascend up a certain whiteness, in the first place, which demonstrates all the virtues of the first ENS of Sulphur; and therefore may we deservedly account of it the first ENS of Sulphur, and so term it. Having thus written of the four first Entities in general, 'tis to be further noted, viz. in what manner they are to be made use of, that their virtues may be perceived, the which is thus: Each of these first Entities is to be put into good wine, in such a quantity that it may be tinged therewith; which done, 'tis prepared for this regiment, or work. Of this wine must you drink every day in the morning about day-break; so long, till your nails of your fingers first fall off, and then the nails on your feet, then your hair and teeth; and then lastly, till your skin be dried up, and new be again generated.

All this being done, you must cease from that Medicament, or Potion, so there will new nails be born again, new hairs, new teeth, and withal, a new skin; & the diseases both of the body and mind will depart away, as is afore—mentioned. Herewith we'll conclude this, our small book of RENOVATION and RESTAURATION.

By Ph. Theophrastus, Bombast of Hohenheim, a Philosopher, a Monarch, a Spagyrical Prince, a Most great Astronomer, a wonderful Physician and Trismegist of Mechanick Arcanas.

A Word from the Publisher

Thank you for purchasing this small work from The R.A.M.S. Library of Alchemy. During his lifetime, Hans Nintzel was dedicated to the identification, acquisition, study, retyping and, when necessary, translation of what he considered to be the most important known works on Alchemy. Hans was assisted by his sparse network of fellow Alchemists, all members of the Restorers of Alchemical Manuscripts Society (R.A.M.S.). I was an active member of R.A.M.S.

My goal is to publish all of the works originally made available through R.A.M.S. as photocopies. To facilitate this, I have chosen to have the books professionally printed. I also have a few titles that I intend to add to the original R.A.M.S. Library, selected by strict criteria established by Hans.

The works from the original R.A.M.S. Library are republished by R.A.M.S. Publishing Company in the collection, "The R.A.M.S. Library of Alchemy," with permission of the Estate of Hans W. Nintzel.

If you have a work on Alchemy that you believe should be a part of the R.A.M.S. Library, please contact me through R.A.M.S. Publishing Company.

Philip N. Wheeler